the history of life

THE PRENTICE-HALL FOUNDATIONS OF EARTH SCIENCE SERIES

A. Lee McAlester, Editor

STRUCTURE OF THE EARTH
S. P. Clark, Jr.

EARTH MATERIALS
W. G. Ernst

THE SURFACE OF THE EARTH
A. L. Bloom

EARTH RESOURCES, 2nd ed.
B. J. Skinner

GEOLOGIC TIME, 2nd ed.
D. L. Eicher

ANCIENT ENVIRONMENTS
L. F. Laporte

THE HISTORY OF THE EARTH'S CRUST*
A. L. McAlester and D. L. Eicher

THE HISTORY OF LIFE, 2nd ed.
A. L. McAlester

OCEANS, 2nd ed.
K. K. Turekian

MAN AND THE OCEAN
B. J. Skinner and K. K. Turekian

ATMOSPHERES
R. M. Goody and J. C. G. Walker

WEATHER
L. J. Battan

THE SOLAR SYSTEM*
J. A. Wood

*In preparation

the history of life

second edition

A. LEE McALESTER

Dean, School of Humanities and Sciences,
Southern Methodist University

PRENTICE-HALL, INC., *Englewood Cliffs, New Jersey 07632*

Library of Congress Cataloging in Publication Data

McALESTER, ARCIE LEE, (date)
  The history of life.

  (The Prentice-Hall foundations of earth science series)
    Bibliography: p. 159
    Includes index.
    1. Evolution. 2. Paleontology. I. Title.
QH366.M24 1977   575   76-24905
ISBN 0-13-390146-7
ISBN 0-13-390120-3 pbk.

© 1977, 1968 by Prentice-Hall, Inc.

Englewood Cliffs, New Jersey 07632

20 19 18 17 16 15 14 13 12

Printed in the United States of America

PRENTICE-HALL INTERNATIONAL, INC., *London*
PRENTICE-HALL OF AUSTRALIA PTY. LIMITED, *Sydney*
PRENTICE-HALL OF CANADA, LTD., *Toronto*
PRENTICE-HALL OF INDIA PRIVATE LIMITED, *New Delhi*
PRENTICE-HALL OF JAPAN, INC., *Tokyo*
PRENTICE-HALL OF SOUTHEAST ASIA PTE. LTD., *Singapore*
WHITEHALL BOOKS LIMITED, *Wellington, New Zealand*

contents

introduction   1

one

the beginnings of life   3

CHEMICAL MODELS OF THE ORIGIN OF LIFE, *4*

STRATEGIES OF LIVING SYSTEMS, *9*

PRECAMBRIAN LIFE, *13*

THE EUCARYOTE EXPANSION, *19*

two

the diversification of life   28

EARLY CONCEPTS, *29*

DARWIN AND NATURAL SELECTION, *30*

THE RISE OF GENETICS, *33*

MODERN EVOLUTIONARY THEORY, *37*

EVOLUTION AND THE FOSSIL RECORD, *42*

three

life in the sea **44**

PATTERNS OF LIFE IN THE SEA, *45*

MARINE FOSSILS AND FOSSILIZATION, *52*

MARINE PLANTS, *56*

EARLY PALEOZOIC INVERTEBRATE LIFE, *59*

MODERNIZATION OF MARINE INVERTEBRATES, *65*

POSSIBLE CAUSES FOR WIDESPREAD EXTINCTIONS, *68*

four

the transition to land **69**

PLANT TRANSITIONS, *70*

TERRESTRIAL INVERTEBRATES, *72*

ORIGIN OF THE VERTEBRATES, *75*

THE EVOLUTIONARY HISTORY OF FISHES, *78*

AMPHIBIANS—TRANSITIONAL VERTEBRATES, *85*

five

land plants **89**

FOSSILIZATION OF LAND PLANTS, *90*

SEEDLESS FLORAS OF THE PALEOZOIC ERA, *92*

EARLY SEED-BEARING PLANTS, *100*

THE RISE AND DOMINANCE OF FLOWERING PLANTS, *105*

six

reptiles and mammals **108**

EARLY REPTILES, *109*

DINOSAURS AND OTHER RULING REPTILES, *116*

BIRDS, *126*

MAMMAL ORIGINS, *126*

MAMMAL DIVERSIFICATION, *132*

seven

man   137

PRIMATE ORIGINS AND ADAPTATIONS, 138
PRE-PLEISTOCENE MONKEYS AND APES, 143
THE PLEISTOCENE EXPANSION OF MAN, 146
HOMO, 150
THE EVOLUTION OF HUMAN CULTURE, 153

appendix: a classification of organisms   155

suggestions for further reading   159

index   161

geologic time scale   168

introduction

Beginning students often wonder why the history of life is discussed in books and courses on Earth science. After all, isn't *biology* the science concerned with life, whereas Earth science is the study of nonliving phenomena—rocks and minerals, rivers and glaciers, mountains and oceans? Although it may at first seem paradoxical, life history has always been closely linked with Earth science because the only direct evidence for understanding past life is provided by *fossils*, the remains of ancient animals and plants preserved in sedimentary rocks. Geologists, rather than biologists, have been most directly involved with studies of sedimentary rocks and have therefore discovered most of the fossils. As a result, *paleontology*, the science concerned with the study of fossils, has developed over the past hundred years as an aspect of sedimentary geology. The association of paleontology with geology was reinforced by the early nineteenth-century discovery that fossils occur in a definite sequence, which enabled paleontologists to determine the relative ages of fossil-bearing rocks. This discovery was of extraordinary importance, for it quickly led to the establishment of a worldwide geologic time scale, yet it impeded an understanding of the history of life because paleontologists tended to become interested in fossils only as tools for dating rocks, rather than as documents of life history. Biologists, on the other hand, have always been concerned with the origins of present-day animals and plants, but have seldom tested their conclusions by examining the fossil record. As a consequence, there has traditionally been little interchange between biology and the geologically oriented science of paleontology.

Fortunately, this long and unnatural separation is breaking down, for modern paleontologists are becoming increasingly interested in comparative studies of living organisms as a means of understanding life of the past. At the same time, biologists are beginning to appreciate the unique value of fossils as primary documents of evolutionary history. The essential point is that geological and biological evidence are both necessary for understanding the historical development of life on Earth. Both kinds of evidence will be discussed in this book.

The organization of the book is generally chronological; it begins with the first primitive organisms of several billion years ago and ends with the rise of human civilization and the beginnings of written history about 5,000 years ago. This chronological treatment, which stresses the evolutionary sequence of fossils, is supplemented by discussions of two related aspects of the study of fossils in other volumes of this series. The uses of fossils as indicators of geologic age are treated in *Geologic Time*, 2nd edition, by D. L. Eicher; the many applications of fossils to understanding past environments are discussed in *Ancient Environments* by L. F. Laporte.

Most books about life history, particularly those that stress the fossil record, require the memorization of long lists of names, both for the fossils themselves and for their many anatomical features. The real excitement of life history lies not, however, in sterile nomenclature, but in more fundamental problems and concepts. For this reason, technical terms will be introduced only where they are really essential for understanding the subject. There are, however, two groups of terms that *are* essential: the chronologic age terms of the *geologic time chart* and the names of the *principal groups of animals and plants*. The geologic time chart has been reproduced on the last page of the book. Students not already familiar with it should memorize it as soon as possible. A complete reference classification of animals and plants is given in the Appendix beginning on page 155. Students with little or no background in biology may wish to supplement the classification by additional readings from some of the general references listed in Suggestions for Further Reading, p. 159.

As we progress through the chronology of life, it will become apparent that there is a relationship between the time chart and the principal groups of organisms, for life has become increasingly diverse and complicated through time. The less complex animals and plants therefore tend to be dominant in earlier geologic periods and the more advanced organisms in later periods of geologic time. We shall also see that ancient organisms were surprisingly similar to present-day life, for most early forms have persisted even when overshadowed by their more advanced descendants. In spite of such well-known exceptions as dinosaurs and trilobites, most of the principal kinds of life that have *ever* arisen are still living on Earth today. Far from being a strange progression of bizarre and exotic creatures, the history of life is a record of the persistent survival of familiar organisms found around us in our modern world.

one

the beginnings of life

Fossilized remains of organisms are first preserved abundantly in rocks about half a billion years old. The Earth itself is much older, having consolidated from cosmic dust at least $4\frac{1}{2}$ billion years ago. Sometime during the 4 billion years that made up the first eight-ninths of our planet's history, there was an obscure but extraordinarily significant event—the beginning of life on Earth. Most biologists now believe that life originated by a process of slow development from nonliving chemical systems; today few areas of science offer more challenge and excitement than does the search for understanding the nature of this development and the primitive organisms that it produced. Since the mid-1950's, research in two disparate fields has added significantly to our knowledge of these earliest organisms. Biochemists have become interested in laboratory syntheses of the chemical components of life under conditions that simulate those of the primeval Earth; and geologists have discovered fossilized remains of early life in *Precambrian* rocks (those formed during the eight-ninths of Earth history before the Cambrian Period, which marks the beginning of abundant fossilized life). In this chapter we shall review these developments by considering the experimental evidence for the origin of life from nonliving chemical systems and the life processes of the simplest organisms that exist today. Then we shall turn to the documentary evidence for early life provided by Precambrian fossils.

CHEMICAL MODELS OF THE ORIGIN OF LIFE

Like the initial differentiation of the Earth itself into a core, mantle, crust, ocean, and atmosphere, the origin of life on Earth probably occurred in the dim interval of earth history that antedates the oldest surviving rocks. This is apparent because the remains of ancient organisms are found in sedimentary rocks of almost all ages, including some formed over 3,000 million years ago, and thus the beginnings of life must have taken place still earlier. Because we have no direct evidence of this first transition between nonliving and living systems, we are forced to rely on indirect clues provided by speculations and experiments on the probable chemical environment of the early Earth.

The Components of Life

Most of us think of the living world as extraordinarily complex and diverse. The differences between an oak tree and an earthworm, or between a chimpanzee and a bacterium, are so striking that it is hard to imagine that all life is made up of only a few essential types of chemical compounds. Chemically, however, organisms are but endlessly varied permutations of five principal constituents: *water, carbohydrates, fats, proteins,* and *nucleic acids* (Fig. 1-1).

Compounds	Functions	Composition
Water	Universal solvent	Hydrogen, oxygen
Carbohydrates	Energy source	Hydrogen, oxygen, carbon
Fats	Energy storage	Hydrogen, oxygen, carbon
Proteins	Structural; facilitation of chemical reactions	Hydrogen, oxygen, carbon, nitrogen, phosphorus, sulfur (organized into 20 amino acid "building blocks")
Nucleic acids (DNA, RNA)	Patterns for protein construction	Hydrogen, oxygen, carbon, nitrogen, phosphorus (organized into 5 nucleotide "building blocks")

FIG. 1-1   Principal chemical constituents of life.

Water is by far the most common of these constituents. It is a solvent for many substances within organisms, and it also enters directly into the molecular structure of other essential compounds. Carbohydrates and fats serve primarily as energy suppliers. Carbohydrates, such as sugar and starches, are readily soluble in water and thus provide a quick source of energy; fats, on the other hand, are relatively insoluble and serve principally as a means of long-term energy storage. The remaining two groups of compounds, the proteins and nucleic acids, are perhaps the most fundamental of all. Most structural elements in organisms (for example—cell membranes, muscles, or hair) are composed of proteins, as are enzymes, the essential substances that facilitate chemical reactions in all living systems. Proteins, in turn, are large and complex molecules made up of linear chains of smaller molecular units, called *amino acids*, of which there are about 20 types. Diverse combinations of these 20 amino acid building blocks account for the great variety of protein structure and function. Nucleic acids, the final life component, are intimately related to proteins. In the more advanced kinds of animals and plants, they are concentrated in the cell nucleus where they provide templates which serve as patterns for the manufacture of all the different proteins required by the cell. Nucleic acids are themselves made up of linear chains of five *nucleotide bases*. Differing combinations of these five bases provide a code for the manufacture and linkage of amino acids to form proteins.

All these major chemical constituents of life, and most minor constituents as well, are compounds dominated by four elements—hydrogen, oxygen, carbon, and nitrogen. Carbon, in particular, is fundamental because of unique chemical properties that permit it to link the other elements into the large and complex molecules characteristic of organisms. The first requirement for the origin of life, then, was a supply of these chemical elements in forms that might combine into proteins, nucleic acids, and the other components of living systems. There are suggestions that just such conditions may have existed early in the history of the Earth.

Environment of the Early Earth

Spectroscopic analysis of sunlight, combined with chemical analyses of meteorites, provide means of calculating the relative abundance of the chemical elements in the Solar System. These calculations show that the four elements essential for life are among the most abundant in the Solar System today (Fig. 1-2). Hydrogen is first in abundance, oxygen third, carbon fourth, and nitrogen

FIG. 1-2 The most abundant chemical elements in the solar system. The four principal elements of organisms are shaded.

Rank	Element	(Standard Symbol)	Relative Abundance (Silicon = 100)
1	Hydrogen	(H)	2,500,000
2	Helium	(He)	380,000
3	Oxygen	(O)	2,500
4	Carbon	(C)	930
5	Nitrogen	(N)	240
6	Silicon	(Si)	100
7	Magnesium	(Mg)	91
8	Neon	(Ne)	80

fifth. Because the Earth, Sun, and planets probably originated from a cosmic dust cloud of more or less uniform composition, the four elements may have been abundant on the newly formed Earth, just as they are today throughout much of the Solar System. Indeed, they were probably more abundant very early in Earth history than at any time since, for there are strong indications that the Earth has lost most of its original complement of the lighter and more volatile elements of the Solar System.

Today, the unreactive rare gas elements (argon, neon, krypton, and xenon) are millions of times less abundant on the Earth than in the Sun. Similarly, hydrogen, carbon, and nitrogen, elements that most commonly occur as gases or in gaseous compounds, are also thousands of times less abundant today on the Earth and terrestrial planets than in the Sun and outer planets. These facts indicate that, early in its history, the Earth somehow lost its original solar gases, and that its subsequent atmosphere and ocean have been secondarily derived by volcanic release of gases trapped within the solid Earth.

The earliest secondary atmosphere released from within the solid Earth was also almost certainly dominated by light gaseous elements, particularly hydrogen, carbon, and nitrogen, but their exact chemical form is uncertain. Present-day volcanic gases are largely carbon dioxide with smaller amounts of water vapor, nitrogen, hydrogen, methane (CH_4), and other gases. Likewise, the atmospheres of Venus and Mars, the Earth's neighboring terrestrial planets, are today dominated by carbon dioxide. These facts have led most geologists to conclude that the Earth's earliest atmosphere most probably had a similar composition. Such a primitive atmosphere is, however, less compatible with the chemical synthesis of life than would be one that contained carbon combined only with hydrogen as methane gas (CH_4), rather than with oxygen as carbon dioxide (CO_2). Numerous laboratory experiments (described in the next section) show that the chemical precursors of life form far more easily in environments that lack oxygen-containing compounds other than water. For this reason, some geologists have postulated a very early stage in the Earth's volcanic release of gases, during which oxygen was relatively deficient and hydrogen compounds, such as methane (CH_4) and ammonia (NH_3), were dominant. Other than its greater probability as an environment for the origin of life, there is no direct evidence for such an early atmosphere.

"Primitive Earth" Experiments

It is, of course, a long step from an atmosphere containing any form of hydrogen, carbon, oxygen, and nitrogen to even the simplest living systems, but, at the least, atmospheric evidence suggests that the principal elements now found in organisms were also abundant in some form on the early Earth. This supposition was about as far as one could reasonably go until 1953, when S. L. Miller, then a graduate student at the University of Chicago, performed a now-classic experiment on the origin of life. Miller constructed an apparatus for circulating steam through a mixture of ammonia, methane, and hydrogen (Fig. 1-3). The steam-gas mixture was subjected to a high-energy electrical spark and then condensed to a liquid before the cycle was begun again by heating the water to steam. After a week of being cycled through the gases and electrical discharges, the condensed water in the apparatus had become deep red and turbid. On analysis, the water was discovered to contain a complex mixture of amino acids, the basic structural units of proteins. Miller's simple experiment showed that electrical discharges, such as lightning, in an early atmosphere of ammonia, methane, and hydrogen could have led to the production of some of the complex molecules of living systems, and this discovery opened a new field of investigation—the experimental synthesis of the chemical constituents of life under "primitive Earth conditions."

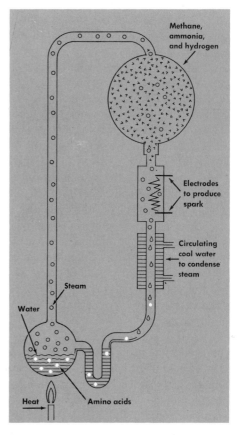

FIG. 1-3 Miller's apparatus for producing the organic building blocks of life from a simulated "primitive atmosphere" of methane (CH_4), ammonia (NH_3), and hydrogen(H_2). When exposed to a high-energy electrical spark simulating natural lightning amino acids are formed.

Since 1953, many such experiments have been performed with various gaseous mixtures and energy sources that might have been present early in Earth history. These experiments have employed such gases as carbon dioxide, water vapor, methane, nitrogen, ammonia, and hydrogen with various possible energy sources, such as strong solar radiation, lightning, thunderlike sound waves, and meteorite shock waves, any of which might have caused the relatively simple atmospheric compounds to combine into larger and more complex molecules. Although none of these experiments has produced anything even approaching the complexity of the simplest organism, they *have* succeeded in showing that a variety of the complex chemical building blocks that make up life could have been present in the early ocean and atmosphere. In addition to amino acids, both carbohydrates and nucleotides (the structural units of nucleic acids) have been shown to form when any of the possible energy sources is applied to various gaseous mixtures containing the elements hydrogen, carbon, nitrogen, and oxygen. The only limitation on the synthesis of these materials is that they are formed far less readily when oxygen-containing compounds such as carbon dioxide dominate the gaseous mixtures. Principally for this reason, some scientists have postulated that the earliest atmosphere was dominated by hydrogen, methane, and ammonia.

These studies have led scientists to visualize a time early in Earth history when the surface was covered with oceans or lakes that were rich in molecules fundamental to life. The waters of these oceans or lakes have been often described as a "dilute organic soup," a concept first developed in the 1920's and 1930's by the English biologist J.B.S. Haldane and the Russian biochemist A. I. Oparin, pioneer workers on the origin of life. In the burst of interest following Miller's experiment in 1953, the ideas of Haldane and Oparin have been greatly expanded, and a number of speculative hypotheses now attempt to explain the development of the first self-duplicating organisms from the non-living building blocks of the early organic soup.

A principal difficulty in these hypotheses lies in explaining how relatively small and uniform amino acid molecules became linked into much larger and more complex proteins. Such linkage seldom takes place spontaneously in the presence of water, but it does occur when amino acids become partially dried. For this reason, some workers have stressed the importance of such periodically dried environments as beaches or tide pools in providing sites for the synthesis of larger molecules.

A still more fundamental difficulty involves the organization of such molecules, once formed, into discrete, organized cells capable of self-reproduction. Here it is known that under certain rather restricted conditions, organic compounds suspended in water tend to aggregate spontaneously into small spheres bounded by a wall or membrane, which separates this material from the more dilute surrounding liquid. How such spherical concentrations of organic compounds might have developed the extraordinarily complex, self-duplicating chemical systems seen in even the simplest living cell remains, however, a fundamental puzzle.

STRATEGIES OF LIVING SYSTEMS

Because there is no direct experimental evidence of the transition between complex carbon molecules and actual living cells, perhaps the best glimpses of early life are provided by the simplest organisms that still exist today. The two most fundamental life processes of all living systems are, first, the gathering of *structural materials* necessary to construct and repair the organism, and, second, the utilization of some external *energy source* for organizing these materials.

Structural Materials

All organisms must have a continuous supply of the basic components of life—carbon, hydrogen, oxygen, and nitrogen—for essential structural proteins and enzymes. There are, however, two fundamentally different ways in which living systems obtain these essential materials (Fig. 1-4).

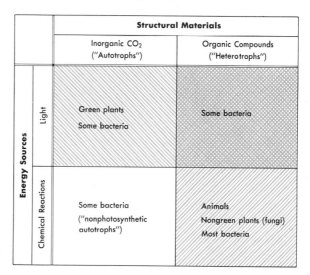

FIG. 1-4 The principal sources of structural materials and energy in present-day life.

Many organisms, called *autotrophs*, can synthesize structural materials directly from atmospheric carbon dioxide and water. The most familiar present-day autotrophs are green plants, which utilize solar energy to convert water and atmospheric carbon dioxide into carboyhdrates through a complex series of reactions that are known collectively as *photosynthesis*. The carbohydrates so produced are then utilized as a secondary energy source for the manufacture of structural proteins, the necessary nitrogen being supplied as dissolved nitrogen compounds in the water utilized by the plant. (This, of course, is the reason that nitrogen is a principal component in fertilizers used to stimulate plant growth.) Although most autotrophs obtain nitrogen from nitrogen compounds dissolved in water, a few, called *nitrogen-fixers,* are able to utilize gaseous nitrogen from the atmosphere for the production of proteins. These autotrophs have the simplest possible nutritional requirements, for they can subsist only on carbon dioxide and nitrogen from the atmosphere, plus water. Included in this group are certain microscopic bacteria, as well as some larger, seaweedlike plants.

In addition to this diverse array of autotrophs, there is today an equally diverse spectrum of organisms called *heterotrophs.* These cannot utilize atmospheric carbon dioxide as a carbon source, but are, instead, dependent on autotroph-produced organic molecules, principally proteins, for their structural materials. All animal life, as well as most bacteria and such nongreen plants as fungi, fall into this category.

At first glance, the earliest life might seem to have been simple autotrophs, perhaps similar to some present-day photosynthetic bacteria that manufacture all their components from atmospheric carbon dioxide and nitrogen. Biochemists are convinced, however, that the chemistry of photosynthesis is far too complex to have arisen in the earliest living systems. It thus appears more probable that, instead of directly synthesizing organic materials, the first organisms were heterotrophs that depended on nonbiological carbon compounds for structural materials. In other words, they must have "eaten" the organic soup from which they arose.

Energy

In addition to sources of essential carbon and nitrogen, all organisms require sources of energy for the utilization of these structural materials. Because such complex carbon compounds as carbohydrates, fats, and proteins are also rich in potential chemical energy, they are used *both* as structural materials and energy sources by most heterotrophs. Conversely, green plants, the most familiar autotrophs, use the energy from sunlight to produce carbohydrates, which in turn supply secondary chemical energy back to the plants themselves and to organisms that may later eat the plants.

The principal energy sources of present-day life are sunlight and carbon compounds, yet certain bacteria show less familiar sources that might have played a larger role in the early history of life. Some of these bacteria use the chemical energy trapped in various *inorganic* compounds of sulfur and nitrogen for the production of structural materials from atmospheric carbon dioxide; these are called *nonphotosynthetic autotrophs* (Fig. 1-4). Such forms have the disadvantage of being dependent on a continuous supply of the energy-giving compounds, but have the strong advantage of being able to live within muds, soils, and other environments where sunlight cannot penetrate.

In contrast, still other bacteria are heterotrophs, deriving structural carbon from organic compounds, yet using sunlight as an energy source for the breakdown of the compounds. In short, present-day bacteria show every possible combination of sources for their structural materials and energy. It should be stressed, however, that most modern bacteria use organic compounds both as structural materials and energy sources, just as do all present-day animals.

Procaryotes and Eucaryotes

The previous discussion has emphasized the unique adaptive versatility of microscopic, single-celled organisms called bacteria. With their close relatives, the seaweedlike *blue-green algae*, bacteria also show differences in cell structure and function that sharply separate them from all other organisms. So fundamental are these differences that they serve to divide present-day life into two great subcategories, whose distinctions are far more basic than are the familiar differences between the plant and animal "kingdoms" (see the Appendix, A Classification of Organisms, p. 155).

Bacteria and blue-green algae, collectively known as *procaryotes*, have cells that lack a chromosome-bearing nucleus to transmit messages for cell construction during reproduction. All other organisms, from single-celled plants and animals to oak trees and elephants, are made up of cells having such a nucleus. These nucleus-bearing organisms are collectively known as *eucaryotes*.

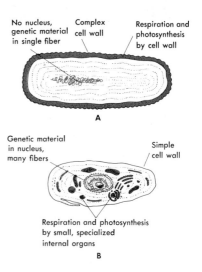

FIG. 1-5 (A) Procaryotic cell; (B) eucaryotic cell.

In addition to the presence or absence of a cell nucleus, procaryotes and eucaryotes differ in many other basic structural features, some of which are summarized in Fig. 1-5. These differences indicate that procaryotes have, in general, a less complex overall makeup than do eucaryotes. This further suggests that the earliest life may have had procaryotic structure and that eucaryotes arose from procaryote ancestors. We shall see later that there is independent geologic and biologic evidence to support this conclusion, but first we shall

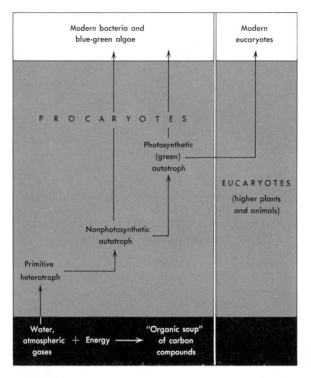

FIG. 1-6 Inferred stages in the earliest history of life.

review the probable sequence of events in early life history suggested so far (Fig. 1-6).

Before the first organisms arose, molecules of amino acids, carbohydrates, and other carbon compounds accumulated in the surface waters of the Earth through the action of solar energy, lightning, and perhaps other energy sources acting on gases of the primitive atmosphere, such as methane, ammonia, or (somewhat less probably) carbon dioxide. These molecules became aggregated into larger proteins and nucleic acids. By processes that are still uncertain, they developed into the first self-duplicating organisms, which were probably simple heterotrophs using preformed carbon compounds as both nutrients and energy sources. Gradually these organisms developed the ability to synthesize complex nutrients from simpler compounds, and ultimately the first autotrophs arose, which may have been similar to some modern procaryotic bacteria that require only atmospheric carbon dioxide and nitrogen as nutrients. Like some present-day bacteria, these earliest autotrophs may have used the chemical energy of inorganic minerals, rather than sunlight, as an energy source, for such a scheme of chemical synthesis is less complex than is the process of photosynthesis. From these first "nonphotosynthetic" autotrophs, developed the first green plants, which may have been similar to certain modern photosynthetic bacteria. From these, in turn, could have arisen both more complex procaryotic bacteria and blue-green algae, as well as the host of advanced eucaryotic organisms.

PRECAMBRIAN LIFE

Studies of the chemistry and life strategies of present-day organisms provide many clues to events in the early history of life. In contrast, the only *direct* evidence of early life is provided by the study of fossils, the remains of past organisms that became trapped and preserved in ancient sedimentary rocks.

Fossils are not equally abundant throughout the Earth's long sedimentary record; most are found in rocks deposited only during the past 600 million years. This abundant fossil record provides a universal means of dating and subdividing this more recent, or *Phanerozoic*, interval of Earth history. Older rocks, spanning the long Precambrian interval from about 600 to 4,000 million years ago, have far fewer and less conspicuous fossil remains. Indeed, until rather recently it was believed that Precambrian sedimentary rocks contained *no* unequivocal indications of ancient life. Since the mid-1950's, however, it has become clear that simple procaryotes, both bacteria and blue-green algae, were abundant in certain favorable environments throughout most of the long Precambrian interval of Earth history.

Chemical Traces of Early Life

Precambrian sedimentary rocks contain two fundamentally different kinds of evidence of early life. The first are mere chemical traces of the activities of life; the second are the actual preserved remains of the organisms themselves.

The most common chemical traces of early life are provided by complex carbon compounds extracted from Precambrian sediments. Recent development of very precise analytical techniques has led to the identification of many such compounds, but there are two difficulties in their interpretation. (1) It is possible that the compounds did not originate at the same time as the rocks, but seeped into the sediments long after they were deposited (or were formed in place by later bacterial activity). (2) It is difficult to prove that the compounds are actually the products of organisms, for, as we have noted, nonbiological carbon compounds—making a sort of organic soup—were probably common in the waters of the early Earth.

The first problem—time of origin of the compounds—has proved difficult to solve, yet there is good evidence that many of the extracted compounds did, at *some time*, result from life processes. Large protein or carbohydrate molecules could be almost certain evidence of life processes, but such complex molecules are unstable and tend to decompose to simpler carbon compounds when buried for long periods. These decomposition products often resemble nonbiological carbon compounds. Geochemists are convinced, however, that many of the extracted compounds, although relatively simple in structure, are nevertheless too similar to organism-produced materials to have formed nonbiologically. Further evidence of biological origin is provided by a distinctive property of the carbon compounds manufactured by organisms.

FIG. 1-7 Precambrian stromatolites, layered, hemispherical masses of lime-stone deposited by the metabolic activities of blue-green algae. These illustrations show two-billion-year-old specimens from Great Slave Lake, Canada. (Above) Top view of many hemispherical masses. (Below) Horizontal cut through several masses, showing the layered structure. (Courtesy Paul Hoffman and Geological Survey of Canada.)

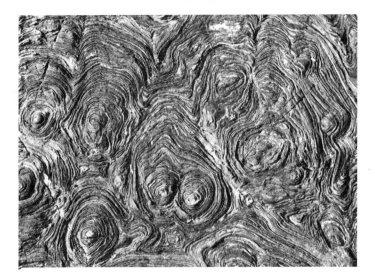

Naturally occurring carbon consists of a mixture of two principal nuclear species, or *nuclides*, which differ in weight. They are known as carbon-12 and carbon-13. Carbon-13 has an additional neutron in the nucleus and is thus slightly heavier than carbon-12. When autotrophic green plants remove carbon dioxide from the atmosphere, they tend to selectively concentrate the lighter nuclide in the carbohydrates that they produce. These carbohydrates, in turn, supply nutrients to most of the living world, with the result that life-produced carbon compounds today always show a somewhat lower proportion of heavy carbon-13 than do nonliving carbon compounds. Most significantly, most of the carbon extracted from Precambrian sediments also shows this selective depletion of carbon-13, which strongly suggests that it was initially produced by some sort of autotrophic, most probably photosynthetic, life.

A second and more direct kind of chemical trace of Precambrian life occurs as layered masses of limestone or chert, known as *stromatolites*. Similar layered deposits also occur in Phanerozoic sedimentary rocks and are being formed today in shallow, warm oceans by the metabolic activities of blue-green algae. Because of their resemblance to these younger deposits, Precambrian stromatolites have long been suspected as being the metabolic remains of Precambrian blue-green algae (Fig. 1-7). Somewhat similar materials, however, can also originate by direct precipitation of silica or calcium carbonate without the intervention of life, and, for this reason, the algal origin of Precambrian stromatolites was long a subject of debate. This uncertainty has been resolved over the past two decades by the discovery of the actual preserved remains of ancient algae and bacteria in many Precambrian stromatolites. These discoveries have revolutionized our understanding of the living world through the long Precambrian interval of Earth history.

Fossil Procaryotes

Scattered reports of microscopic fossils from Precambrian rocks have been made for at least 50 years, but it was not until 1954 that the first convincing discovery was made of ancient fossil procaryotes. They were found in the Gunflint Chert, a formation of Precambrian stromatolites about 1,900 million years old exposed along the shores of Lake Superior (Fig. 1-8). Since 1954, several dozen additional localities, ranging in age from about 600 to over 3,000 million years, have also yielded Precambrian procaryotes, yet the original Gunflint discovery remains one of the most diverse and well-preserved records of these early fossils.

Gunflint Chert Fossils. The Gunflint Chert is exposed in northern Minnesota and adjacent Ontario as a part of a series of Precambrian sedimentary rocks about 50 feet thick, overlying an older volcanic and granitic terrain. The fossil-bearing unit occurs in the middle of the sedimentary sequence as a 10-foot layer, made up of alternating thin layers of white, red, and black cherts (rocks

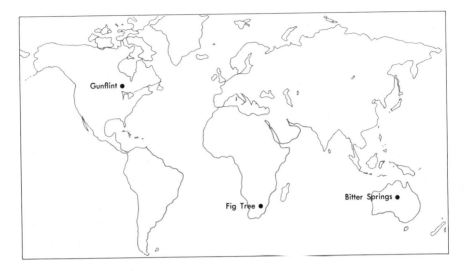

FIG. 1-8 Important fossil-bearing exposures of Precambrian rocks.

composed of very fine-grained silica) and iron-bearing carbonates. The cherts occur in dome-shaped, stromatolitic masses, several feet in diameter, that, like similar Precambrian stromatolites, long have been supposed to be algal deposits. In 1954, S. A. Tyler, a field geologist, and E. S. Barghoorn, a paleobotanist, dissolved some of the chert in hydrofluoric acid and discovered a residue of tiny filaments and spheres that were obviously fragments of primitive organisms. Soon after death, the fragments apparently became embedded in gelatinlike silica, which must have hardened rapidly into dense chert that protected the delicate structures from extensive deformation and decay. Most Precambrian procaryotes discovered since have been found in similar cherts. They are best studied by examining very thin slices of the chert under high magnification with optical and electron microscopes; such studies have shown the Gunflint organisms to fall into three broad categories (Fig. 1-9):

1. Thin threads, some with wall-like partitions, that closely resemble present-day filamentous bacteria and blue-green algae (Fig. 1-9(A)).

2. Spherical bodies, probably of diverse origin. Some resemble present-day bacteria and unicellular blue-green algae; others may be reproductive spores of blue-green algae; still others are unlike any present-day organism (Fig. 1-9(B)).

3. Star-shaped, umbrella-shaped, or parachute-shaped bodies of unknown affinities (Fig. 1-9(C)).

The similarities of the first two kinds of structure to present-day procaryotic bacteria and blue-green algae are extremely important, for these are precisely the kinds of organisms one would expect to find early in the history of life. Although some modern bacteria are autotrophs, most are not, deriving their nutrients instead as parasites or from the decay of dead organisms. Modern blue-green algae are all photosynthetic autotrophs, however, and it is most probable that many of the Gunflint fossils also derived their nutrients by photosynthesis from atmospheric carbon dioxide and solar energy. Radioactive age determinations show the rocks of the Gunflint Chert to be about 1,900 million years old, and the process of photosynthesis was therefore almost certainly developed by that time.

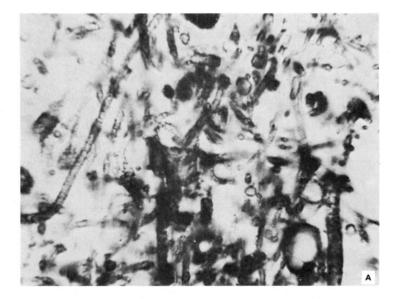

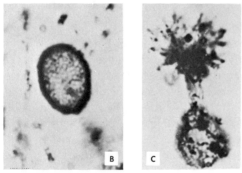

FIG. 1-9 Procaryote remains from the 1,900 million year old Gunflint Chert of Ontario, Canada, magnified from 1,000 to 2,000 times. (A) Thread-shaped forms which closely resemble modern filamentous bacteria and blue-green algae. (B) Spherical form resembling a modern bacterium. (C) Parachute-shaped form of unknown affinities. (Courtesy Elso Barghoorn.)

Fig Tree Group Fossils.   Because the Gunflint Chert fossils include rather advanced photosynthetic procaryotes, great interest attaches to any older Precambrian fossils that might reflect still earlier stages of procaryotic development. Unfortunately, only a few fossil-bearing localities have been found in rocks significantly older than the Gunflint; of these, perhaps the most important occur in the Fig Tree Group of South Africa, a thick series of sedimentary shales, sandstones, and cherts exposed in the Barberton Mountains of the Transvaal Region (Fig. 1-8). Radiometric dating shows the Fig Tree sediments to have been deposited more than 3,200 million years ago, which makes them among the oldest sedimentary rocks yet discovered.

As with the Gunflint fossils, the microscopic Fig Tree organisms are preserved in cherts. They are, however, far less diverse and well preserved than are the Gunflint fossils, as might be expected because of their much greater age. Two principal forms occur (Fig. 1-10):

1. Rod-shaped forms whose size, shape, and cell wall structure suggest that they are bacterial remains (Fig. 1-10(A)).

2. Spherical structures with a granular surface that somewhat resemble certain simple blue-green algae in size and shape (Fig. 1-10(B)).

A

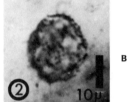

B

FIG. 1-10 Probable procaryote remains from the Fig Tree Group of South Africa, magnified from 1,000 to 2,000 times. These fossils, over 3,200 million years old, are among the oldest preserved remains of life. (A) Rod-shaped forms resembling bacteria. (B) Spherical forms with a granular surface, possibly a simple blue-green alga.

The presence of bacterialike procaryotes so early in Earth history is not surprising, but if the spherical Fig Tree structures are indeed the remains of blue-green algae—a conclusion that is still uncertain—then photosynthetic autotrophs must have been present from near the beginning of the Precambrian sedimentary record. If so, the origin of photosynthesis, like the origin of life itself, must have occurred in the dim pregeological interval of Earth history. It is also significant that the oldest known stromatolites, which were almost certainly formed by blue-green algae, were deposited about 2,800 million years ago, and thus are only slightly younger than the Fig Tree microfossils. This too indicates that algal photosynthesis developed very early in the preserved record of life.

Bitter Springs Formation Fossils. In addition to the Gunflint Chert, several dozen younger Precambrian localities have now yielded preserved microfossils, and perhaps the most significant of these occur in the Bitter Springs Formation of central Australia (Fig. 1-8). This formation consists of limestones, sandstones, and cherts deposited only about 900 million years ago, not long before the sudden expansion of animal life that began the Phanerozoic Era, about 600 million years ago.

Like the Gunflint Chert, the Bitter Springs Formation contains well-preserved procaryotic bacteria and blue-green algae preserved in cherts. As might be expected, these show a more modern aspect than do the much older Gunflint fossils. In particular, the blue-green algae are more diverse and show closer similarities to still living representatives of the group. Several forms, in fact, are practically indistinguishable from certain present-day blue-green algae, which makes these the longest lived of any modern organism.

The Bitter Springs fossils also include small spheres that contain what, on first inspection, appear to be the degraded remains of cell nuclei. Such structures would be of great importance because only eucaryotic cells have nuclei. These fossils, which have the general size and shape of certain eucaryotic green algae living today, were first accepted as the oldest eucaryotes. Recent restudy shows,* however, that the supposed cell nuclei are in fact the shrunken remains of the entire original cell, rather than degraded nuclei. It thus appears that the Bitter Springs deposits, as well as all older Precambrian rocks, lack fossil eucaryotes.

THE EUCARYOTE EXPANSION

The first unequivocal record of fossil eucaryotes begins about 700 million years ago with the worldwide appearance of large, multicelled fossil plants and animals (Fig. 1-11). Many aspects of this critical juncture in life history are still obscure, yet enough facts are known to permit some tentative conclusions about its nature and possible causes.

Eucaryotes, Metaphytes, and Metazoans

Viewed in somewhat greater detail, the expansion of eucaryotic life presents two separate but interrelated puzzles. First, how did the earliest eucaryotes originate from procaryotic ancestors? Second, why did advanced multicellular eucaryotes have such a sudden and dramatic appearance about 700 million years ago?

There is little direct fossil evidence bearing on the first question—manner of origin—yet the subject has been a source of much controversial speculation among biologists in recent years. This has mostly involved the ultimate source of two kinds of small, spherical substructures found only within eucaryotic cells. These structures, called *mitochondria* and *chloroplasts*, are both involved in eucaryote energy transfer; chloroplasts are the pigmented sites of photosynthesis in green plants, and mitochondria are the sites of foodstuff oxidation where simple energy-giving compounds are made available to the rest of the cell. In

---

*A. H. Knoll and E. S. Barghoorn, *Science*, 190, October 3, 1975, 52–54.

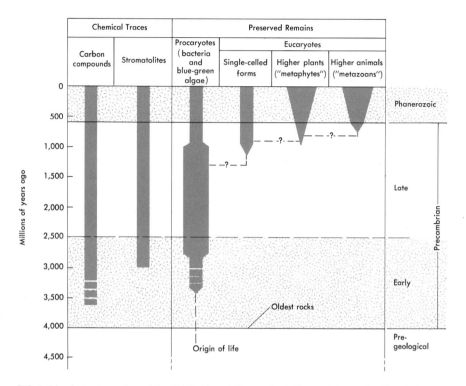

FIG. 1-11 Summary review of the distribution of life remains in Precambrian rocks. Chemical traces of ancient life as well as fossil procaryotes (bacteria and blue-green algae) are found through much of the Precambrian record, but eucaryotic remains occur only in late Precambrian deposits.

FIG. 1-12 Two hypotheses for the origin of the eucaryotic cell from a procaryote ancestor. Specialized internal structures (mitochondria and chloroplasts) might have originated from either (A) infolding of the cell wall, or (B) beneficial invasion of the larger cell by smaller procaryotes.

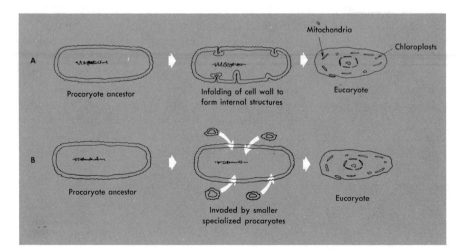

procaryotes, both of these functions are carried out by the complex cell wall rather than by separate internal structures. It is thus conceivable that both mitochondria and chloroplasts developed from specialized, budlike outgrowths of the procaryotic wall (Fig. 1-12(A)). Rather surprisingly, however, it has also been discovered that both structures contain their own separate complement of gene-bearing nucleic acids, which, on cell division, are transferred to offspring cells independent of the principal genetic material contained in the cell nucleus. This has suggested to several workers that mitochondria and chloroplasts were once independent organisms that invaded a simple procaryote to produce the earliest eucaryote (Fig. 1-12(B)). According to this theory, then, eucaryotes arose not from an internal modification of procaryotic structure, but from a beneficial or "symbiotic" invasion of certain simple procaryotes by others. Both ideas, it should be noted, have difficulty explaining the single most distinguishing feature of eucaryotes—the concentration of most of the cell's genetic material into a chromosome-bearing nucleus.

Regardless of their manner of origin, the earliest eucaryotes were most probably inconspicuous, single-celled forms similar to certain simple green algae* that still exist today. Such forms, in turn, probably gave rise to a host of microscopic, single-celled eucaryotic plants and animals in late Precambrian time, which, in turn, developed into larger, multicelled plants and animals (Fig. 1-11). To distinguish them from their smaller, single-celled relatives, such multicelled eucaryotes are known as *metaphytes* ("changed plants") and *metazoans* ("changed animals").

Unfortunately, there is almost no fossil documentation of evolutionary transitions from single-celled eucaryotes to either metaphytes or metazoans; both abruptly appear fully developed in latest Precambrian rocks and have been abundant fossils ever since. The reason for this sudden appearance of large metaphytes and metazoans about 700 million years ago remains a second fundamental puzzle of life history, which we shall consider after reviewing the earliest fossil record of multicellular life.

The Ediacara Fauna, the Oldest Animals

Scattered and rather poorly preserved remains of both metaphytes and metazoans are found in latest Precambrian rocks from almost every continent, but the first really good look at advanced eucaryotic life is provided by a remarkable association of metazoan fossils from Australia. These were first discovered in 1947 when an Australian geologist found some rounded impressions resembling fossil jellyfish in sandstones of the Ediacara Hills in South Australia. These sandstones were presumed to be of Cambrian age, and because such

---

*See "A classification of organisms," p. 155. Note that only one algal group, the blue-green algae, are procaryotes. All other algae, as well as all higher plants and animals, are eucaryotes.

impressions are relatively common in Cambrian rocks, the discovery passed without notice. Several years later, two private collectors discovered not only jellyfish but also impressions of segmented worms and other, more problematic organisms in the same rocks. These new fossils, which did not resemble any known Cambrian organisms, prompted an intensive restudy of the Ediacara material by M. F. Glaessner, a paleontologist at the University of Adelaide. Glaessner's work soon showed that the strange fossils occurred in a continuous sequence of undeformed sedimentary rocks, in which the only other fossils were typical Early Cambrian forms, and these were found about 500 feet stratigraphically above the problematic fauna. Because of the distinctiveness of the Ediacara fossils and their occurrence below typical Cambrian forms, it seems highly probable that they represent latest Precambrian organisms, with an estimated age of 650 to 700 million years.

Several thousand specimens have so far been collected from the Ediacara Hills, mostly rather large organisms one to several inches long. All are preserved as impressions of the original animal along bedding planes of sandstone. Six principal types have been discovered:

1. Rounded impressions, some with radiating grooves, that resemble modern jellyfish (primitive animals of the Phylum Coelenterata, see p. 156). These forms are the only fossils in the Ediacara fauna that are also common in younger rocks (Fig. 1-13(A)).

2. Impressions of stalklike fronds with grooved branches that suggest living "sea pens," primitive colonial marine animals that also belong to the Phylum Coelenterata (Fig. 1-13(A)). Similar but less well-preserved impressions have also been reported from late Precambrian rocks of Southwest Africa, England, and elsewhere.

3. Elongate, wormlike impressions consisting of a horseshoe-shaped head followed by about 40 identical segments. These impressions almost certainly represent animals similar to living segmented worms of the advanced Phylum Annelida (Fig. 1-13(B)).

4. Rounded, flattened, wormlike impressions with a central groove and strong segmentation. Traces of an intestinal tract have been reported, suggesting a relationship to somewhat similar-looking living worms of the Phylum Annelida (Fig. 1-13(C)).

5. Oval, shield-shaped impressions with T-shaped grooves; these do not closely resemble any known organism, although it has been suggested that they may be primitive representatives of the Phylum Arthropoda (Fig. 1-13(D)).

6. Circular impressions with three "bent arms" radiating from the center; these also resemble no known organism (Fig. 1-13(E)).

FIG. 1-13 (Facing page) The oldest known fossil animals, from the Late Precambrian Ediacara Hills of South Australia. The fossils are preserved as impressions in sandstone. (A) Rounded, jellyfishlike forms (lower left), and branched, sea penlike form (upper right) (one-third natural size). (B) Elongate, wormlike form (magnified 2 times). (C) Rounded, wormlike form (actual size). (D) Oval, shield-shaped form, of unknown affinities (magnified 2 times). (E) Circular form with "bent arms," of unknown affinities (magnified $2\frac{1}{2}$ times). (Courtesy M. F. Glaessner.)

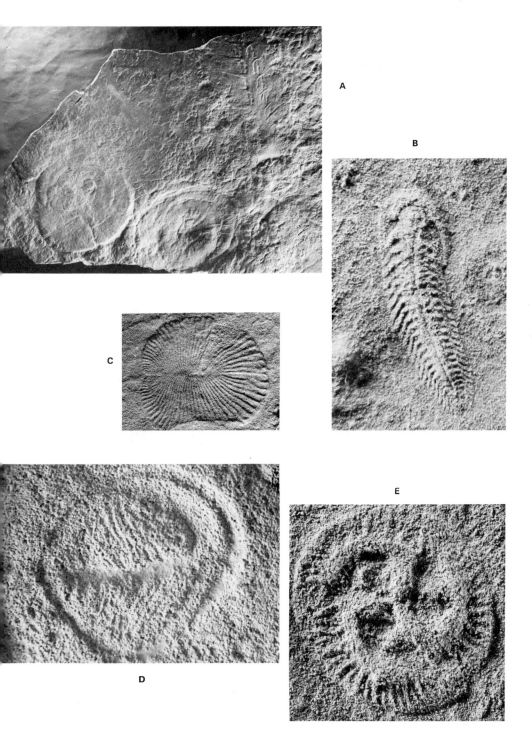

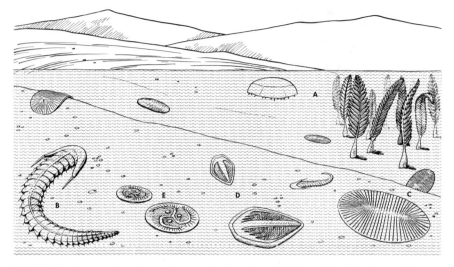

FIG. 1-14 Reconstruction of the Ediacara animals as they might have looked in a Precambrian sea. The letters correspond to the photographs in Fig. 1-13.

The Ediacara fossils thus include two unequivocal present-day animal phyla: (1) the *Coelenterata*, a phylum of primitive multicellular organisms that includes modern corals, jellyfish, and sea anemones; and (2) the *Annelida*, a more advanced phylum of segmented worms that includes the modern earthworms. Figure 1-14 is a reconstruction showing how these oldest known fossil animals might have looked in their life environment.

These earliest fossil metazoans, although relatively simple and unspecialized, were nevertheless enormously more complex than were their presumed single-celled eucaryotic ancestors. Not only are the Ediacara animals many-celled creatures with highly differentiated systems of muscles, nerve cells, and food-gathering organs, but they also most probably had fully developed sexual reproduction as do all present-day metazoans. Thus, they mark dramatic evolutionary advances that, regrettably, are not yet documented by still earlier Precambrian fossils, which might show transitional stages between these highly developed animals and simpler eucaryotes.

Early Cambrian Life and the End of the Eon

The Ediacara fauna is made up entirely of soft-bodied, wormlike, and jellyfishlike animals preserved as mere impressions in sandstone. Such preservation requires rather special conditions of burial and original sediment chemistry, and thus fossil remains of soft-bodied life are rare compared with more easily preserved animal "hard parts," such as shells or skeletons (see p. 63). Apparently shell or skeleton-bearing animals had not yet evolved when the Ediacara sediments were deposited about 600 to 700 million years ago, for such remains are everywhere absent from rocks of that age or older. The Ediacara animals, however, clearly foreshadow what is probably the most dramatic single event in all of life's history—the rapid expansion of shell-bearing animals that ends the long Precambrian interval of the Earth's past. Such animals appear abundantly and almost simultaneously in rocks about 550 to 600 million years old found today on every continent, and these occurrences mark the beginning of Cambrian time and the Phanerozoic Eon, which will be our subject throughout the remainder of the book.

The early Cambrian expansion of animal life includes not just one or two new adaptations; instead, early representatives of almost all of the major types of shell-bearing marine animals that are still found in the ocean today occur in early Cambrian sedimentary rocks. In Chapter 3 we shall look more closely at these principal types of marine animals and trace their history through the Cambrian and succeeding periods. For the present we need only emphasize that the early Cambrian fossil faunas are dominated by many-legged trilobites, which, although now extinct, were distant relatives of modern crabs, and by brachiopods, two-shelled, superficially clamlike animals that are common today in some tropical oceans (Fig. 1-15). In addition, early relatives are found of such dominant modern groups as sponges, clams, snails, and starfish.

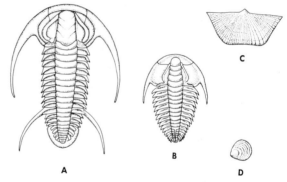

FIG. 1-15 Typical trilobites (A, B) and brachiopods (C, D) are the most abundant animal fossils of early Cambrian rocks.

The Ediacara soft-bodied metazoans and the subsequent dramatic expansion of shell-bearing metazoans in Cambrian time bring us back to the second of our major questions concerning the expansion of eucaryotic life—why did advanced metazoans rather suddenly appear and expand only about 700 million years ago?

Until rather recently it was possible to speculate that soft-bodied metazoan animals, such as those of the Ediacara fauna, might have actually been common through much of Late Precambrian time, but were not preserved as fossils because of their lack of shells or skeletons. It is now known, however, that even soft-bodied metazoans leave distinctive patterns of trails and burrows in the muds and sands of the ocean floor in which they live. Recent studies of such animal-produced structures in Late Precambrian sedimentary rocks show that they first appeared about 700 million years ago, at about the time of the Ediacara fossils. Thus, the late and abrupt appearance of metazoan animals is almost certainly a real and fundamental puzzle of life history.

Many speculative suggestions, mostly falling into two broad categories, have been advanced to explain this puzzle. One group of hypotheses relates the rapid metazoan expansion to some facilitating change in the physical world, such as more benign climates or changes in the chemistry of the oceans or atmosphere. The other suggestions emphasize direct changes in the organization or interactions of the animals and plants themselves, rather than changes in some external, physical cause. No hypothesis from either category has been widely accepted, but we shall briefly review one recent suggestion of each type.

A suggestion of the second, nonphysical sort relates the rise of both multicellular animals and plants to the development of sexual reproduction. Simple one-celled organisms, whether procaryotes or eucaryotes, can reproduce rather easily by duplication of critical cell components, followed by division of the cell into two individuals. More complex, multicellular eucaryotes, particularly plants, sometimes reproduce similarly by "budding," in which a part of the parent expands and breaks off to form a new individual. More typically, however, multicellular eucaryotes have sexual reproduction in which special germ cells from different individuals, the eggs and sperm, unite to form a new individual. As we shall see in Chapter 2, such reproduction offers enormously greater prospects for modifying the genetic material, structure, and adaptations of the offspring than does simple asexual division. For this reason, some workers have suggested that the rapid expansion of multicellular animals and plants in Late Precambrian time simply marks the time when eucaryotic life first developed the complicated mechanisms of sexual reproduction.

A more complex hypothesis that has been widely discussed in recent years relates the metazoan expansion to changing concentrations of atmospheric oxygen.

We have noted that the Earth's early atmosphere was probably made up largely of carbon dioxide, water vapor, and nitrogen released from melting and volcanic outgassing of the solid earth. The present atmosphere also contains carbon dioxide, water vapor, and nitrogen, but differs profoundly from the early atmosphere in that it also contains large quantities of gaseous oxygen. Most of this oxygen has apparently been released into the atmosphere through photosynthesis by green plants.

Some geologists now believe that the great expansion of metazoan life that ends Precambrian time reflects the fact that the atmosphere first accumulated abundant plant-produced oxygen at about that time. Almost all modern metazoans require free oxygen for their life processes. In contrast, many present-day bacteria and blue-green algae, modern descendants of the procaryotes that were present through much of Precambrian time, do not require free oxygen, either in the atmosphere or dissolved in the water surrounding them. Photosynthesis by blue-green algae, however, releases oxygen as a waste product (bacteria, even though some are photosynthetic, have different biochemical processes that do not release oxygen).

Initially, any oxygen released by blue-green algae would have quickly combined with dissolved iron and other oxygen-deficient elements and compounds in the surrounding waters, rather than being released to the atmosphere. Eventually, however, most of these elements would have become oxidized by the algae-produced oxygen. Then free oxygen would begin to accumulate slowly in the oceans and bubble to the ocean surface to be released into the atmosphere. The development and expansion of oxygen-dependent metazoan life late in Precambrian time may thus mark the time when the atmosphere and oceans first accumulated abundant free oxygen by this process. If so, then during much of Precambrian time only oxygen-independent, procaryotic life would have been possible on Earth. Nevertheless, the life processes of procaryotic blue-green algae, acting over hundreds of millions of years, could have slowly released oxygen to the atmosphere, and thus paved the way for the dramatic expansion of oxygen-dependent life that brings to an end the long Precambrian phase of Earth history.

two

the diversification of life

Diversification of life had already taken place before fossils became abundant in Early Cambrian time, for the procaryotes and advanced invertebrate animals found in Precambrian rocks are far more complex and varied than were the still more primitive organisms from which they arose. Yet the panorama of life's diversity was only beginning to unfold, for ahead lay the sequence of changes that were to transform tiny aquatic plants into towering sequoia trees, and small wormlike animals into huge reptiles, soaring birds, swift-footed mammals, and, ultimately, into man himself. Unlike the inconspicuous record of life before the Cambrian Period, abundant fossils document these later stages in the history of life. These stages will be our concern in following chapters, but in this chapter we will interrupt our chronological discussion to consider the mechanisms that have enabled organisms to increase in diversity and complexity—in short, the topic of *organic evolution*. This chapter provides a brief introduction to *how* and *why* evolutionary changes have taken place in organisms. In the chapters that follow we will be considering *what* changes have occurred, and *when* and *where* they took place.

EARLY CONCEPTS

Man has always been concerned with the living things he finds around him. Aboriginal hunters, Greek philosophers, medieval monks, and modern outdoorsmen have all noticed that individual animals and plants, although differing from one another in detail, tend to fall into natural groupings of similar individuals. These groupings are called *species*. Thus, all individual peach trees combine to make a single species, all domestic dogs another, and all edible American oysters still a third. Following a scheme first proposed by the great eighteenth-century Swedish botanist Carolus Linnaeus (1707–1778), scientists have customarily given each living and fossil species a distinctive two-part Latin name. The peach tree is *Prunus persica*, the dog *Canis familiaris*, and the oyster *Crassostrea virginica*. The second word in each name (*persica, familiaris, virginica*) distinguishes the species; the first word (*Prunus, Canis, Crassostrea*) applies to groupings of similar species called *genera* (singular, *genus*). Thus, the genus for peach trees, *Prunus*, also includes the plum tree, *Prunus domestica*, and other closely related species. Similarly, the genus *Canis* includes other doglike animals such as the coyote, *Canis latrans; Crassostrea* includes, among other species, the Portuguese oyster, *Crassostrea angulata*, and the Japanese oyster, *Crassostrea gigas*. Just as related species are combined into genera, so are similar genera combined into *families*, families into *orders*, orders into *classes*, and classes into *phyla* (singular, *phylum*), the highest category. All of the phyla and many of the classes of organisms, as well as selected examples of lower categories, are given in the Appendix, A Classification of Organisms, beginning on page 155. Note, however, that only two categories, genus and species, are reflected in the formal Latin name of each species.

Species were already being named and described in the early eighteenth century, but until the middle of the nineteenth century naturalists generally agreed that each species had been separately created, either by the action of a Supreme Being or by "spontaneous generation," a vague concept according to which fully developed organisms were believed to have sprung from water, soil, or other nonliving matter. The idea that the more complex animals and plants might have developed by gradual change from simpler forms had been suggested in Classical times, but it did not again receive serious attention until about 1800, when it was restated by several progressive English and French naturalists. The most articulate and important of these was a gifted Frenchman, Jean Baptiste de Lamarck (1744–1829). In 1809 Lamarck published a consistent and well-reasoned theory that species are not "immutable," but that they descend instead from other species by gradual change over many generations. Lamarck called this process *transformism*; the word *evolution* was not applied to the process until many years later. Lamarck's work clearly anticipated many fundamental ideas that were later to be popularized by others, but it was his misfortune to be ahead of his time. His theory was strongly criticized by leading naturalists of his day and, as a result, never received the attention or credit it deserved. It remained for Charles Darwin, writing 50 years later, to convince the scientific world of the truth of transformism.

DARWIN AND NATURAL SELECTION

Like most naturalists of his day, Darwin (1809–1882) had often debated the revolutionary idea of transformism (or *transmutation*, as it was by then more commonly called), and, like his contemporaries, he began by disbelieving it (Fig. 2-1). His notebooks show that throughout his early career he steadfastly maintained that each species had been separately created. The turning point came in 1837, a few months after his return to England from a five-year voyage as naturalist aboard the naval survey ship *Beagle*. Reflecting on the experiences of the voyage, it occurred to Darwin that certain puzzling facts about South American fossil mammals and the distribution of living species on isolated islands could best be explained by the heretical idea of transmutation. The more he considered the idea, the more probable it seemed, yet there was one key still missing—an adequate mechanism to explain just how one species changes into another.

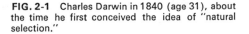

FIG. 2-1 Charles Darwin in 1840 (age 31), about the time he first conceived the idea of "natural selection."

Earlier evolutionists believed that useful changes occurring in animals during their lifetimes were passed on to the next generation. The usual example was the neck of the giraffe, which was thought to have gradually lengthened as the animals stretched to feed on higher and higher tree leaves. In each generation the neck muscles became longer by active use, just as the muscles of an athlete become larger and stronger with exercise. It was assumed that these changes, acquired by an individual, were in some way passed on to his offspring, thus resulting in a longer neck in each generation. This concept of the "inheritance of acquired characters" was particularly stressed in Lamarck's writings, and for that reason it is sometimes referred to as *Lamarckism*. Darwin did not doubt that this process was an important cause of evolutionary change, but he nevertheless felt that some additional mechanism was required to explain the endless diversity of animals and plants.

Two years of intensive reading and thinking were to pass before Darwin devised another explanation for transmutation. Impressed with the wide variation in such features as size, shape, and strength that are found among individuals of the same species, and realizing that most organisms produce many offspring that die before maturity, he reasoned that the individuals surviving to reproduce the species must be those with the most successful combinations of variable characters. Thus, only the "fittest" individuals pass their desirable variations on to the next generation. To use Darwin's own example,

> Let us take the case of a wolf, which preys on various animals . . . and let us suppose that the fleetest prey, a deer for instance, had . . . increased in numbers, or that other prey had decreased in numbers, during that season of the year when the wolf was hardest pressed for food. I can under such circumstances see no reason to doubt that the swiftest and slimmest wolves would have the best chance of surviving and so be preserved or selected . . . I can see no more reason to doubt this, than that man can improve the fleetness of his greyhounds by careful and methodical selection.

Over many generations this selective reproduction by the most successful individuals would lead to adaptive changes in the species and, ultimately, to new species. Darwin called this process *natural selection* to distinguish it from the man-made, or artificial selection practiced by breeders of domesticated animals and plants.

Always a thorough and cautious worker, Darwin set about gathering evidence to convince the scientific world of the truth of transmutation and natural selection. This work occupied him, along with other writings, for 20 years and might have lasted longer had not another English naturalist, Alfred Russel Wallace, independently discovered natural selection in 1858. Wallace wrote to Darwin about his new theory and, in a rare example of scholarly statesmanship, the two men agreed to publish their independent discovery simultaneously as two short papers. These papers attracted little attention, but Wallace's discovery stimulated Darwin to complete the longer treatise that had been in preparation since 1837. In 1859 he published *The Origin of Species by Means of Natural Selection, or The Preservation of Favoured Races in the Struggle for Life*, a work that was to become one of the most influential books ever written (Fig. 2-2).

The 500 pages of the *Origin* brought together, in an informal and sometimes rambling style, a wealth of evidence for both the fact of evolution and the mechanism of natural selection. The crux of the argument is developed in the first four chapters, "Variation under Domestication," "Variation under Nature," "Struggle for Existence," and "Natural Selection, or the Survival of the Fittest." The remaining 11 chapters are extended commentaries on the advan-

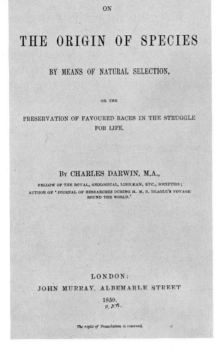

ON

THE ORIGIN OF SPECIES

BY MEANS OF NATURAL SELECTION,

OR THE

PRESERVATION OF FAVOURED RACES IN THE STRUGGLE
FOR LIFE.

By CHARLES DARWIN, M.A.,

FELLOW OF THE ROYAL, GEOLOGICAL, LINNÆAN, ETC., SOCIETIES;
AUTHOR OF 'JOURNAL OF RESEARCHES DURING H. M. S. BEAGLE'S VOYAGE
ROUND THE WORLD.'

LONDON:
JOHN MURRAY, ALBEMARLE STREET
1859.

The right of Translation is reserved.

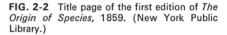

FIG. 2-2 Title page of the first edition of *The Origin of Species*, 1859. (New York Public Library.)

tages and difficulties of natural selection as an explanation for evolutionary change. Such factors as hybridism, instinct, the fossil record, geographical distribution, and embryology are considered and shown to be consistent with the theory. The book closes with an effective and often-quoted passage of Victorian prose:

> Thus, from the war of nature, from famine and death, the most exalted object which we are capable of conceiving, namely, the production of higher animals, directly follows. There is grandeur in this view of life, with its several powers, having been originally breathed by the Creator into a few forms or into one; and that, whilst this planet has gone cycling on according to the fixed law of gravity, from so simple a beginning endless forms most beautiful and most wonderful have been, and are being evolved.

The impact of the *Origin* was immediate and sensational. The literary public had already been introduced to the notion of evolution by popularized accounts published before 1859. That an established and respected scientist was now attempting to prove the revolutionary idea seemed to capture the popular imagination, with the result that the entire first edition of the book was sold on the first day of publication. Increasing unrest with the idea of "special creation," coupled with Darwin's persuasive mass of evidence, led many influential scientists to become immediate converts to his views. Although there were some strong opponents, opinion was so generally favorable that by 1868, only nine years after the publication of the *Origin*, Darwin was able to write truthfully of the "now almost universal belief" in evolution.

THE RISE OF GENETICS

It is a tribute to Darwin's genius that his explanation of evolutionary change by natural selection is still the cornerstone of modern thinking about evolution. Natural selection has not, however, been continuously esteemed as an evolutionary process since Darwin's time, for early in this century Darwin's ideas were completely overshadowed by equally fundamental discoveries that were made in the new science of genetics.

The principal contribution of genetics to the understanding of evolution has been the explanation of the inheritance of variability in individual organisms of the same species, for, as we have seen, it is this variability that provides the raw materials for natural selection and the rise of new species. The basic principles of inheritance were first discovered in 1865 by an obscure monk named Gregor Mendel (1822–1884), who lived in Moravia, a part of modern Czechoslovakia. Unfortunately, Mendel's work was unknown to Darwin and was, indeed, completely ignored until its essence was rediscovered by other workers around 1900. From 1900 to 1925 knowledge of heredity rapidly developed and, at the same time, Darwin came to be thought of as old-fashioned because of his incorrect ideas about the inheritance of individual variations. We shall see that it is only in the years since about 1925 that natural selection has again been recognized as an essential part of the evolutionary process.

Genes and Reproduction

The work of Mendel and later geneticists has shown that the development of the individual organism is controlled by hereditary regulators known as *genes*. Genes are constructed of the nucleic acid called DNA (deoxyribonucleic acid) and are normally located in the cell nucleus, where they are organized into larger, paired, threadlike units called chromosomes, each of which may contain thousands of genes. The number of chromosomes is usually constant for each species but varies between species, ranging from as few as one pair to as many as several hundred pairs. The usual number is between 5 and 30 pairs. Man, for example, has 23 pairs. When a cell divides during normal growth, the chromosomes reproduce themselves exactly to give, in the two new cells, the same number and kinds of chromosomes as in the original parent cell. This process of exact chromosome duplication is called *mitosis*. In most organisms with sexual reproduction, a more specialized kind of cell division called *meiosis* takes place in the organs where *gametes* (specialized reproductive cells such as eggs and sperm) are produced.

There are two cell divisions in this second process. The first is similar to mitosis in producing two new cells, each with a complete complement of paired chromosomes. These two cells further divide in a manner that leaves only one set of the original paired chromosomes in each of the four offspring cells (Fig. 2-3). These become the gametes, each of which carries half the number of

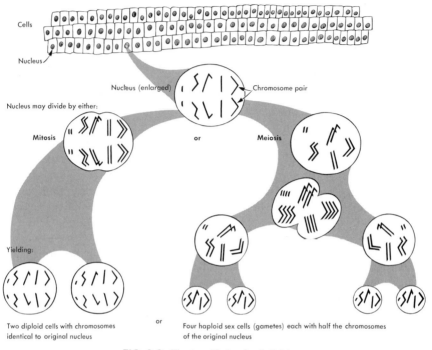

Cells

Nucleus

Nucleus (enlarged)   Chromosome pair

Nucleus may divide by either:

Mitosis   or   Meiosis

Yielding:

Two diploid cells with chromosomes   or   Four haploid sex cells (gametes) each with half the chromosomes
identical to original nucleus   of the original nucleus

FIG. 2-3   The two kinds of cell division.

chromosomes necessary for the final organism. When two gametes meet in the process of fertilization, a new organism is produced, one that receives half of its chromosomes from each of the two parents. Thus, meiosis and subsequent fertilization provide a means of interchanging genetic material between organisms, whereas mitosis provides a means of exactly duplicating cells within an individual organism. In man, all cell divisions are achieved by mitosis except those in the testes of the male and ovaries of the female, where sperm and eggs are produced by meiosis.

The chromosomes are the larger units of heredity, but it is the smaller genes that ultimately determine the nature of the individual organism. Genes may exist in different expressions, called *alleles*, each of which leads to a different hereditary result in the adult organism. In man, for example, blue eyes are the result of one eye-color allele, whereas brown eyes result from another. Recall that chromosomes normally occur in pairs. It is a fundamental fact of heredity that chromosomes that pair have comparable sets of genes that control the development of the same structures in the adult organism. Cells produced by mitosis ther fore have two complete sets of genes, one set contained in each member of the homologous chromosome pairs. Such cells are called *diploid* cells. In meiosis, on the other hand, only one member of each chromosome pair is transmitted to the daughter cell, which thus has only half as many chromosomes as the diploid parent cell. Such cells are called *haploid* cells. Haploid cells contain *one complete set* of genetic instructions, whereas diploid cells contain *two complete sets*. Diploid cells, with their double set of chromosomes, may have the same or different alleles for a particular gene in each set. In man, for example, an individual may have the blue-eye allele in both chromsomes of the pair that contains the eye-color gene, or he may have the brown-eye allele in both chromosomes, or, finally, he might have the brown-eye allele in one chromosome and the blue-eye in the other. If both chromosomes contain the same allele, then all gametes produced by dividing the pairs during meiosis will also have the same eye-color allele. Individuals with the same allele on both chromosomes thus produce only one kind of gamete and are said to be *homozygous* for that particular gene. When the two chromosomes of a pair contain different alleles, the gametes produced by dividing the pairs will be mixed; half will contain the blue-eye allele and half the brown-eye allele. In this case the individual is said to be *heterozygous* for that particular gene (Fig. 2-4).

FIG. 2-4 The inheritance of eye color in human beings. The "brown" allele dominates the "blue" in heterozygotes; thus only individuals that are homozygous for the "blue" allele will have blue eyes.

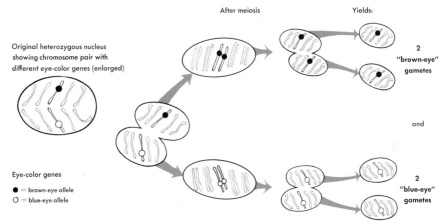

When two gametes unite during fertilization, they will produce either:

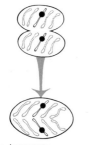

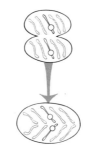

Brown homozygote
(union of 2 "brown-eye" gametes)

Blue homozygote
(union of 2 "blue-eye" gametes)

Heterozygote like parent
(union of 1 "blue-eye"
and 1 "brown-eye" gamete)

You may wonder what color eyes a person will have who is heterozygous for this gene. Often when two different alleles are present, only one, the *dominant* allele, expresses itself, whereas the other, *recessive* allele is passed on to half the gametes, but is not seen in the parent organism. In man, the brown-eye allele is normally dominant over the blue. Note, however, that heterozygous, brown-eyed individuals might have blue-eyed children if they marry partners who are either homozygous or heterozygous for the blue-eye allele. (Why? Could someone who is homozygous for the brown-eye allele have blue-eyed children?)

Our discussion so far has incorrectly implied that each character of the adult organism is controlled by a single gene. Some few characters, such as human eye color, are inherited in this simple way, but most characters are determined by the combined effects of many genes. As the eye-color example shows, predictions about the effects of differing combinations of alleles on the next generation are easily made when only a single gene is responsible for a character, but predictions become increasingly difficult when more genes, each with differing alleles, are involved.

Genes and Individual Variation

With this brief introduction to the fundamentals of inheritance, we can now return to our evolutionary theme and consider the actual causes of the variations found among individuals of the same species. Such variations are of two kinds: those due to heredity and those due to environmental influences operating during the lifetime of the individual. In man, differences in eye color are hereditary variations, whereas the differences in muscle size between an athlete and an office worker of comparable physical build would be an example of variation caused by environmental influences. Only those variations caused by hereditary differences are important in evolution, for they alone can be passed on to the next generation. The children of the athlete will inherit their eye color from him, but they will not have his physique if they choose to become office workers.

Inherited variations, in turn, result from two interrelated processes: mutation and genetic recombination. *Mutation* is the sudden, spontaneous appearance of a new allele for a particular gene or group of genes. Apparently, mutations are continuously taking place in all organisms, but normally they occur at a very low rate. In the fruit fly genus *Drosophila*, which has long been a favorite animal group for genetic study, there is about one gene mutation for every 20 gametes produced. Because each gamete includes about 20,000 genes, the rate of mutation is only about one per 400,000 genes. Mutations may have little or no effect on the adult organism; some, however, are lethal, while others lead to small but advantageous changes. The causes for mutations are obscure, but in most organisms the rate of production can be changed (usually accelerated) by artificial exposure to certain kinds of radiation (gamma, ultraviolet, cosmic), to various chemicals, or to changes in temperature. Apparently, these agents alter the chemical structure of the DNA that makes up the genes, thus producing new alleles.

Mutations are the only source for new alleles but are too rare to be directly responsible for most of the constantly appearing variation found in individuals of the same species. These variations result from simple *recombination*, during meiosis and fertilization, of the alleles already present in the parent organisms. Because the chromosomes of most kinds of organisms contain tens of thousands of genes, each of which may have several alleles, there are almost limitless possibilities for recombination of alleles to produce individuals with differing genetic patterns. It is these differing patterns that lead to the variations seen in adult organisms of the same species.

MODERN EVOLUTIONARY THEORY

During the past 40 years much progress in understanding the process of evolution has resulted from combining Darwin's mechanism of natural selection with the discoveries of geneticists concerning the inheritance of individual variations. A fundamental theme has been the study of inheritance not merely in individual organisms, but in *populations*, which are interbreeding groups of individuals of the same species. It is now recognized that mutation, recombination, and natural selection can lead to evolutionary change only as they act on such groups of individuals; this emphasis has given rise to the science of *population genetics*.

Population Genetics

Population geneticists study changes in gene frequency within interbreeding populations. What was perhaps the most significant discovery concerning such changes was made independently in 1908 by G. Weinberg, a German geneticist, and G. H. Hardy, a British mathematician, and has come to be known as the Hardy-Weinberg law. Hardy and Weinberg demonstrated by simple algebra that the *relative proportion of alleles within a randomly interbreeding population will remain constant unless outside forces work to change it.* Intuitively, one would assume that rare alleles would gradually be lost from the population, and that common alleles would tend to become more common. Instead, Hardy and Weinberg showed that there is a natural *genetic equilibrium*, which can preserve even the least common alleles in a randomly interbreeding population. Only through nonrandom processes, such as mutation or selective reproduction, can the proportion of rare alleles be increased or the proportion of common alleles decreased. This discovery was particularly significant in re-establishing natural selection as an evolutionary mechanism, because, as you will recall, the basis of natural selection *is* nonrandom reproduction. Not every individual, but on the average more of the better fitted, will survive to produce the next generation.

Natural selection is therefore an ideal mechanism for explaining changes in allele frequency and shifts away from genetic equilibrium. Building on the Hardy-Weinberg law, modern population geneticists have developed a body of refined mathematical models, often devised with the help of high-speed computers, to simulate changes in gene frequencies by natural selection in populations that differ in such features as selection pressures, size, original allele ratios, reproductive habits, mutation rates, migration rates, and recombination patterns. In addition to these mathematical formulations, there is now a large body of observational and experimental evidence that confirms the importance of natural selection as a means of changing gene frequencies.

One of the most convincing observations of natural selection in action involves color changes in certain British moths over the past century. These moths live on tree trunks where they were originally protected from predatory birds by their speckled color, which effectively camouflaged them on the lichen-covered trees (Fig. 2-5). Starting about 1850, industrialization in some areas of Britain began to pollute the woods with soot, which killed the lichens and darkened the tree trunks. As the trees darkened, the original speckled moths became more and more conspicuous and were eaten by predatory birds with increasing frequency.

FIG. 2-5 Industrial melanism, an example of natural selection in action. Before the industrial revolution in England, camouflaged, light-colored moths dominated on lichen-covered tree trunks. With the spread of factories in the mid-nineteenth century, dark moths came to dominate on soot-darkened trees. (Left) Light and dark moths on a tree trunk in an unpolluted area. (Right) Light and dark moths on a tree trunk in a polluted woodland near Birmingham, England. (From the experiments of H. B. D. Kettlewell, Oxford University.)

There was, however, in the moth populations a mutant allele that produced a dark gray adult. The dark gray forms were originally quite rare; observations on early collections indicate that they made up fewer than one percent of the moths from the Manchester area in 1848. These rare dark moths were much less conspicuous on the darkening trees and, with increased predation on the normal speckled forms, they began to contribute more and more of the dark-color allele to the next generation. By 1898 the dark forms made up 99 percent of the Manchester populations, and it was now the original speckled moths that were rare. The environmental changes caused by the growth of industry had upset the original genetic equilibrium by altering predation patterns, which, in turn, caused nonrandom survival and reproduction by the better-adapted dark moths. Here, then, was a clear example of natural selection at work. The most successfully adapted individuals were surviving and reproducing, thereby causing an adaptive evolutionary change in the entire population. The original observations on these British moths were made more than 40 years ago and have since been confirmed, both by direct observations of bird predation and by release-recapture experiments using light and dark moths. In addition to the original species studied, about 70 other species of British moths have now been observed to show a darkening trend in industrial regions, and the same phenomenon, which has come to be called *industrial melanism*, has been observed in other areas as well.

These observations provide graphic proof of the power of natural selection to change gene frequencies, but they nevertheless apply only to one or, at most, a very few genes within a single species. Presumably the hundred years or so that man has been carefully observing nature is too short a time for the development of new species or higher categories of organisms, for most directly observed changes have involved only such minor character alterations within populations of a single species (the only exception is the instantaneous origin of new plant species by the process of polyploidy; see p. 41). To understand the more complex processes of species formation, it is therefore necessary to look beyond these directly observable evolutionary changes.

Speciation

Most individual organisms can be rather easily grouped into larger units, called species, that show similar characteristics and that differ more or less markedly from other such groups. Species were being recognized and named long before their evolutionary origin was understood because there usually are clear structural differences between even the most similar species. Even after evolution became an established fact, most biologists were still content to study and describe species without worrying about the processes that originated and maintained their distinctiveness. In the past 40 years or so, however, evolutionary biologists have become increasingly concerned with the process of *speciation*, and today there is a large body of knowledge about the sequence of events leading to the rise of new species.

Interest in speciation was particularly stimulated by the discovery that certain closely similar populations, previously considered members of the same species, *could not interbreed to produce fertile offsping*. The populations were therefore kept from exchanging genes because of subtle barriers to reproduction that were not reflected in the general form and structure of the organisms. This unexpected phenomenon was first noted by population geneticists studying laboratory populations of fruit flies, but it has since been found in wild populations of many kinds of animals and plants. Such reproductively isolated but structurally almost identical populations are called *sibling species*. At about the same time that sibling species were first discovered, biologists studying common, widely distributed species were becoming more and more aware of a complementary phenomenon. Adjacent populations within a species usually show only minor structural differences, but when many populations of a species from many different areas were studied it was usually found that geographically remote or isolated populations differ considerably in form and structure even though they are linked by continuously interbreeding intermediate populations. Species that show such distinctive, geographically separated populations are called *polytypic species*. The separate populations of a polytypic species are called *subspecies*, and are often formally named by adding a third word to the species name. For example, two subspecies of the Horned Owl, *Bubo virginianus*, are *Bubo virginianus subarcticus*, the Arctic Horned Owl, and *Bubo virginianus occidentalis*, the Montana Horned Owl.

The discovery of sibling species, which *look alike but cannot interbreed*, and polytypic species, with subspecies which *look different but can interbreed*, suggested that reproductive barriers preventing gene interchange, rather than simple differences of form, are the really fundamental distinctions between species. As a result, most biologists would now define species as *groups of actually or potentially interbreeding natural populations that are reproductively isolated from other such groups*, rather than merely as groups of structurally similar individuals. Fortunately, in most species reproductive isolation leads to clear differences in form and structure so that they can still be recognized without direct knowledge of their reproductive potentials. It is for this reason that scientists have long been able to describe species with considerable accuracy even without having a clear understanding of the reproductive barriers that maintain their distinctiveness. Modern biologists still rely heavily on structural evidence to recognize species, but today they realize that such evidence is only a convenient and sometimes fallible tool for estimating reproductive isolation. Recognizing species of fossil organisms presents some special difficulties, for, when present-day species are traced back in time, distinctive characters often tend to merge with those of the ancestral species, thus making it difficult to draw sharp structural boundaries between species. In practice, however, the fossil record contains so many gaps in both time and space that paleobiologists can usually recognize structural differences between fossil species that are comparable to those seen between present-day species.

Most biologists believe that the reproductive barriers between species usually originate when relatively small populations become geographically isolated from the parent species for relatively long intervals of time. Natural selection in such isolated populations would, over many generations, lead to differences that would prevent interbreeding and gene flow if the isolated population were subsequently to become geographically reunited with the parent species. Such changes are particularly likely to occur if the isolated population encounters new environments not previously colonized by the parent species. The barriers that prevent interbreeding and gene interchange between species are called *isolating mechanisms* and may be of several sorts. The simplest is complete spatial separation, which of course prevents all contact and gene interchange. Species that are spatially separated from each other, and therefore lack any opportunities to interbreed, are called *allopatric species*; those that come into contact and have opportunities to interbreed are known as *sympatric species*. In allopatric species it is seldom possible to study isolating mechanisms directly, but in closely related sympatric species a number of subtle barriers to interbreeding have been observed. These include: *ecologic barriers*, as when different food requirements or local habitat preferences keep two species from encountering each other even though they live in the same small area; *behavioral barriers*, as when species recognize and prefer only mates of their own species; *structural barriers*, as when changes of shape or form physically prevent reproduction; and *physiological barriers*, as when fertilization or development fail to take place after mating. Most such barriers apparently originate through natural selection when the new species is brought into contact with closely related ancestral species. Often the barriers to reproduction are not fully established, and in such cases *hybridization* may result, leading to intermediate offspring with the characteristics of both parental groups. Normally, hybrids are less well adapted than the parents and either fail to attain sexual maturity or are sterile in the first or second generations. Occasionally, however, hybrids are fertile and have advantageous characters that can lead to still other new species or to the fusion of the two parental groups.

In addition to these relatively slow processes of speciation over many generations, there are mechanisms that can lead to reproductive isolation and speciation in a single generation. The best understood is the process of *polyploidy*, a spontaneous doubling of chromosome number that causes complete reproductive isolation from the parent species. This mechanism was long thought to be of minor evolutionary importance, but it is now known to be a significant means of speciation among higher plants.

Major Adaptive Divergences

We have considered how natural selection operates to change gene frequencies within a single species, and we have also shown how such changes may, over many generations, lead to reproductive isolation and the formation of new species. So far, however, we have said nothing of the possible origins of the major adaptive differences seen in the phyla, classes, orders, families, and genera of organisms. Can the small, relatively subtle changes leading to new species also account for the evolutionary development of the major groups of animals and plants? Most biologists now believe that the processes of mutation, recombination, natural selection, and geographic isolation are sufficient to account not only for the origin of species, but for all evolutionary change. We know, for example, that the amount of potential variability within every organism, each with its thousands of alleles and millions of possible recombination patterns, is far greater than is normally realized because, in a stable environment, natural selection tends to eliminate most combinations and preserve only those that favor stability and continuity of form and habit. There is every reason to believe, however, that in times of rapidly changing environments, there would be strong pressures for natural selection to make fundamental adaptive changes in organisms, changes that may seldom be realized but which are apparently always possible because of the built-in genetic potential of living systems. Consider how effectively natural selection has acted to alter the color of certain moths in response to environmental changes over only the past hundred years. Certainly, then, in the more than 3 billion years since life arose, there have been countless periods of environmental change, each operating over millions of years, that could account for the fundamental differences seen in present-day life. The times involved are so long that we can get few insights from studying only animals and plants living today. Instead, we must turn to the fossil record for a look at these large-scale patterns of evolution.

EVOLUTION AND THE FOSSIL RECORD

The fossil record of animals and plants reveals some surprising evolutionary patterns that would never be suspected from examining evolution as seen only in present-day species. Study of speciation in recent organisms has shown that the changes leading to new species normally take place very slowly over many generations. We might therefore expect that the more complex sequence of changes leading to new genera, families, orders, and classes of organisms would be very gradual indeed. Instead, the fossil record reveals that these fundamental new adaptations tend to arise rather rapidly. Furthermore, the times of origin are not distributed randomly throughout geologic time, but tend to be clustered, so that many new groups appear simultaneously over a short period and then persist with relatively little change for much longer periods. In other words, the rate of evolutionary change in most organisms is not constant but, instead, is extremely variable. Most of the phyla of metazoan animals, for example, originated in Late Precambrian and Early Cambrian time; there have been many evolutionary changes *within* the phyla since Cambrian time, but only one or two new phyla have evolved. Likewise, many of the classes of shell-bearing marine invertebrates that are common in the seas today evolved in Late Cambrian and

Early Ordovician time. Similarly, most of the present-day families of flowering plants originated during the Cretaceous Period, and most of the modern orders of mammals evolved in the Eocene Epoch. These times of rapid diversification are called *evolutionary radiations.*

Other peculiar evolutionary patterns are also seen in the fossil record. One of the most common is the tendency for separate, independently evolving groups of organisms to show the same general sequence of evolutionary changes. Apparently, there are subtle interactions between genetic potential and environmental change that lead different stocks to evolve independently in the same way. This is the phenomenon of evolutionary *parallelism,* or *convergence.* (The term *convergence* is usually applied to distantly related groups that show similar evolutionary tendencies, whereas the term *parallelism* is reserved for closely related groups. The distinction is a subtle one and is not important here.) In extreme cases, separate stocks may become so similar that it is difficult to establish their independent ancestry. Such similar groups descended from more than one ancestral stock are said to be *polyphyletic,* whereas those with a single ancestral stock are termed *monophyletic.* Modern paleontologic study suggests that many major groups may have had a polyphyletic origin. A satisfactory explanation for this repeated tendency toward parallelism and polyphyly is one of the most pressing problems of evolutionary theory.

Not only are the origins of major groups concentrated in time, but so also are their ultimate fates, for the fossil record reveals that organisms have tended to die out simultaneously in relatively sudden, worldwide *extinctions.* Furthermore, the times of extinction are often followed by periods of rapid evolutionary radiation, suggesting that the vacant environments left behind by extinct animals and plants provide an ideal setting for new evolutionary experimentation. The most dramatic extinctions are those near the end of the Permian and Cretaceous Periods that separate the Paleozoic, Mesozoic, and Cenozoic Eras. Smaller extinctions mark the boundaries of the other units of geologic time because, by tradition, the sharp changes in the fossil record that accompany widespread extinctions have been used as criteria for recognizing geologic time units. As with evolutionary radiations, the causes of periodic extinctions are obscure. They are certainly related to environmental changes on the Earth's surface, and, as we resume our chronological survey of life history, we shall consider what some of these changes might have been.

three

life in the sea

In this chapter, and those that follow, we will be considering the enormous evolutionary changes that have occurred during the 600 million years since the beginning of the Cambrian Period. As we continue this chronological survey of life history, we will be concerned mostly with *what* evolutionary changes have taken place, rather than with *how* the mechanisms of natural selection, mutation, recombination, and geographic isolation have led to these changes. Nevertheless, you should keep these essential processes in mind, for they have been responsible for the great diversity of living organisms that we will be discussing.

Unless you happen to live near the seashore, *marine* (sea-dwelling) animals and plants will probably be unfamiliar to you, yet these organisms are of extraordinary importance in understanding the history of life because *most of the phyla of organisms originated in the sea.* If you look over the Classification of Organisms beginning on page 155, you will discover that almost all the phyla are still found in the sea, and also that many phyla, particularly of algae and invertebrate animals, are exclusively marine. Probably all the Precambrian plants and animals discussed in the first chapter lived in the sea, as did all organisms during the Cambrian and Ordovician Periods. It was not until the Silurian and Devonian Periods that the land surface lost its barren and lifeless aspect, for it was then that marine plants and animals first made the transition to land. In this chapter we will survey the nature and history of marine life before describing, in following chapters, the conquest and colonization of the land.

PATTERNS OF LIFE IN THE SEA

Animals and plants living in the sea have three possible modes of life. They can float passively in the water, they can swim actively in the water, or they can live on or within the rocks and sediments of the sea floor. These habits define three great groups of marine organisms: the *plankton* (floating organisms), the *nekton* (swimming organisms), and the *benthos* (bottom-living organisms).

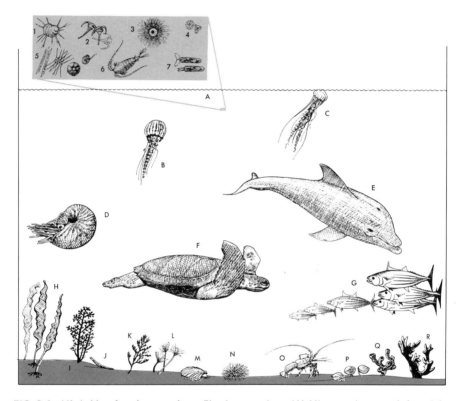

FIG. 3-1 Life habits of marine organisms. *Floating organisms*. (A) Microscopic types: 1, foraminiferan; 2, dinoflagellates; 3, radiolarian; 4, diatoms; 5, coccolithophorids; 6, copepod; 7, pteropods. (B and C) jellyfishes. *Swimming organisms*. (D) cephalopod; (E) mammal (dolphin); (F) reptile (turtle); (g) fishes. *Bottom-living organisms*. (H) brown alga; (I) red alga; (J) lancelet; (K) bryozoan; (L) segmented worm; (M) clam; (N) sea urchin; (O) crustacean; (P) brachiopods; (Q) sponge; (R) coral.

Planktonic organisms are further subdivided into the *phytoplankton* (floating plants) and *zooplankton* (floating animals). There are, of course, no swimming plants so this distinction is unnecessary for the nekton, and benthonic animals and plants are not usually distinguished by special group names. The mode of life of all the principal kinds of animals and plants living in the sea is summarized in Table 3-1 and Fig. 3-1.

Table 3-1   Life Habits of the Principal Phyla of Marine Organisms

	Plants	Invertebrate Animals	Vertebrate Animals
Floating Organisms (Plankton)	PYRROPHYTA (dinoflagellates) CHRYSOPHYTA (diatoms, coccolithophorids)	SARCODINA (foraminiferans, radiolarians) COELENTERATA (jellyfishes) MOLLUSCA (pteropod snails) ARTHROPODA (copepod crustaceans)	
Swimming Organisms (Nekton)		MOLLUSCA (cephalopods)	CHORDATA (fishes, reptiles, mammals)
Bottom-Living Organisms (Benthos)	SCHIZOMYCOPHYTA (bacteria) CHRYSOPHYTA (diatoms) CYANOPHYTA (blue-green algae) CHLOROPHYTA (green algae) PHAEOPHYTA (brown algae) RHODOPHYTA (red algae) MYCOPHYTA (fungi)	SARCODINA (foraminiferans) PORIFERA (sponges) COELENTERATA (corals) BRYOZOA (bryozoans) BRACHIOPODA (brachiopods) MOLLUSCA (snails, clams) ANNELIDA (segmented worms) ARTHROPODA (trilobites, crustaceans) ECHINODERMATA (crinoids, starfish, serpent stars, sea urchins) CHORDATA ("invertebrate chordates"—acorn worms, tunicates, lancelets)	

Plankton

With the principal exception of large floating jellyfish, the plankton is made up of animals and plants that are too small to be easily seen without magnification. Just as on land, almost all life in the sea ultimately depends for nourishment on photosynthesis by green plants. Photosynthesis on land is dominantly the work of grasses, trees, and other relatively large green plants, but in the sea most photosynthesis is accomplished by tiny unicellular green plants, three groups of which—*dinoflagellates, diatoms,* and *coccolithophorids*—are by far the most abundant and important (Figs. 3-1 and 3-2). These tiny forms occur in tremendous numbers in the upper, sunlit layers of the sea. Like their large relatives on land, they use the energy of sunlight to produce carbohydrates from carbon dioxide, in this case carbon dioxide dissolved in the sea. The amount of photosynthesis carried on by these small organisms is staggering to the imagination. It has been estimated that the total bulk of planktonic plant life exceeds that of land plants by a factor of two or three. Most photosynthesis on Earth, therefore, is carried out not by the large green plants that are familiar to us on land, but by microscopic green cells floating in the sea.

There are many more kinds of planktonic animals than of plants, yet the zooplankton, too, is dominated by only a few groups: jellyfish and their relatives; small shrimplike crustaceans known as *copepods*; small floating relatives of the snails called *pteropods*; and tiny unicellular *foraminiferans* and *radiolarians* (Figs. 3-1 and 3-2). Most of these planktonic animals feed directly on phytoplankton and are themselves eaten by a host of larger planktonic, nektonic, and benthonic animals.

Nekton

The dominant marine swimming organisms have always been vertebrates: *fishes, turtles, whales, porpoises, seals,* and various large extinct reptiles. Among the invertebrates, only the *cephalopods* (squids, octopuses, *Nautilus,* and related forms) have become active swimmers in competition with the vertebrates. Most of these swimming animals are active predators feeding on planktonic or benthonic animals, or on each other.

Benthos

Bottom-dwelling life is dominated by large algae, microscopic bacteria and diatoms, a few fungi, and a great diversity of invertebrate animals. The algae, collectively known as *seaweeds,* are confined to relatively shallow, nearshore waters where sunlight can penetrate to the bottom to permit photosynthesis. In shallow waters benthonic algae provide an important source of food and protection for a variety of animals. The total amount of photosynthesis carried out by bottom-living algae is, however, very small in comparison with that of the phytoplankton. Note that almost none of the groups of higher plants that are so familiar to us on land, such as mosses, ferns, and seed-bearing plants, are

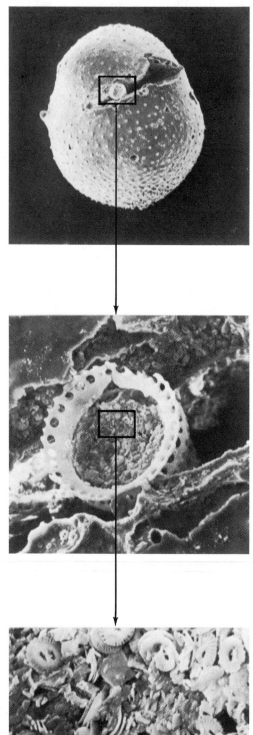

FIG. 3-2 Fossil plankton from Pleistocene sediments. (A) A foraminifer shell about the size of a small pinhead with a still smaller radiolarian jammed in its feeding aperture (magnified 120 times). (B) Close-up view of the radiolarian, showing filling made up of fragments of coccolith shells (magnified 1,000 times). (C) Close-up view of the coccolith shells (magnified 5,000 times). (Reproduced from *Geotimes*, September 1975.)

found in the sea. A few grasslike flowering plants do occur in shallow coastal waters, but they are of minor importance. Instead, it is the various phyla of algae that dominate the sea floor.

Unlike algae, bottom-living animals do not require sunlight, for they can feed on dead and detrital organic matter that falls to the bottom from the sunlit waters above. Thus, the sea floor is almost everywhere inhabited by a diverse fauna of benthonic invertebrate animals. Even the deep ocean basins, whose bottom waters are perpetually dark and cold, support an amazing abundance of invertebrate life. Shallow-water benthonic invertebrates are among the most common fossils in ancient sedimentary rocks because many of them were rather large animals with hard shells or skeletons that were readily preserved, and also because many of them lived in areas of active sedimentation where they were quickly buried. Because of their excellent fossil record, the evolutionary history of bottom-living invertebrate animals is better understood than is that of plank-tonic and nektonic organisms or benthonic plants. For this reason it will be useful to look more closely at the habits and adaptations of benthonic inverte-brates before discussing their evolutionary development.

Adaptations of Benthonic Invertebrates

Bottom-living marine invertebrates eat in one of four different ways. Two of these are the familiar feeding types found in land animals: plant-eating *herbivores*, which in the sea eat mostly the larger bottom-living algae, and *carnivores-scavengers*, which eat other animals, living and dead. The remaining two kinds of feeding are of great importance in the sea, but are rare among land animals. *Filter-feeders*, a large and important group, strain tiny planktonic organisms and detrital organic matter from the bottom water. Most of these animals create currents of sea water by the beat of tiny hairlike structures; the currents are then passed over or through some sort of straining device that traps and accumulates the tiny food particles, often with the help of a sticky mucus, and transfers them to the mouth. The final kind of feeding adaptation is that of the *sediment-* or *deposit-feeders*, which take in organic-rich bottom sediments, utilize part of the organic matter as food, and discharge the undigested sediment particles as feces. Examples of benthonic invertebrates that are adapted for each of these four kinds of feeding are illustrated in Fig. 3-3.

In addition to the four feeding types, bottom-living invertebrates can be divided into four groups on the basis of their living positions and relative mobility. There are *epifaunal* animals that normally live on top of rock or sedi-ment and *infaunal* animals that have adaptations for living within the rock or sediment of the sea floor. Each of these two categories can be further subdivided into two groups. Epifaunal forms can be either attached to the surface or free to move around on the surface. Infaunal forms can be adapted for either bur-rowing into soft sediment bottoms or boring into hard bottoms of rock or wood. Examples of each of these four groups are also shown in Fig. 3-3.

FILTER FEEDERS

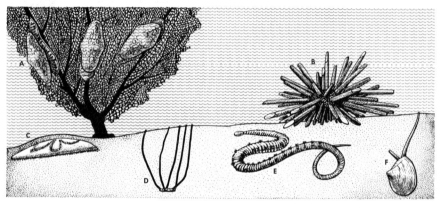

SEDIMENT FEEDERS AND HERBIVORES

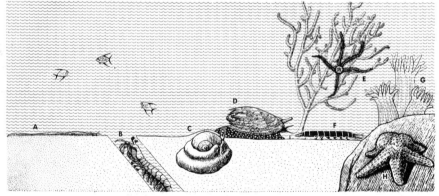

CARNIVORES AND SCAVENGERS

FIG. 3-3 Life habits of bottom-living marine invertebrates. TOP: Filter-feeders. *Mobile:* (K) Crustacea. *Attached:* (A) Crinoidea; (B) Porifera; (C) Bivalvia; (D) Bryozoa; (E) Crustacea; (F) Brachiopoda. *Burrowing:* (I) Annelida; (J) Bivalvia. *Boring:* (G) Bivalvia; (H) Porifera. MIDDLE: *Herbivores (all mobile).* (A) Gastropoda; (B) Echinoidea. *Sediment-feeders (all burrowing).* (C) Echinoidea; (D) Ophiuroidea; (E) Annelida; (F) Bivalvia. BOTTOM: Carnivores and scavengers. *Mobile:* (A) Annelida; (D) Gastropoda; (E) Ophiuroidea; (F) Crustacea; (H) Asteroidea. *Attached:* (G) Coelenterata. *Burrowing:* (B) Crustacea; (C) Gastropoda.

Finally, it is possible to summarize the life habits of bottom-living invertebrates by combining the four feeding types with the four life-position types to give 16 possible life habits (burrowing filter-feeders, mobile herbivores, and so on). The principal groups adapted for each of these habits are shown in Table 3-2. Note particularly that most groups that move around on the surface are either herbivores or carnivores-scavengers, whereas most that are attached to the surface are filter-feeders, except for the corals, many of which use stinging tentacles to capture large prey. Most burrowers in soft sediment are sediment-feeders, although some worms and clams living within the sediment are filter-feeders with special tubes or siphons for bringing in the overlying water for filtration. Likewise, some snails and crustaceans are infaunal carnivores that creep through the sediment in search of buried filter-feeders and sediment-feeders. In addition, only bivalved molluscs and a few sponges have developed adaptations for boring into hard materials such as rock or wood. Note also that 7 of the 16 possible modes of life in the chart are not occupied by *any* animals. As you might expect, there are no epifaunal or boring sediment-feeders, no infaunal or attached herbivores, and no boring carnivores. Finally, note that

Table 3-2 Life Habits of Bottom-Living Marine Invertebrates*

		Filter-Feeders	Sediment-Feeders	Herbivores	Carnivores and Scavengers
Epifaunal (living on the surface of the sea floor)	Mobile	CRUSTACEA		GASTROPODA ECHINOIDEA	GASTROPODA ANNELIDA CRUSTACEA ASTEROIDEA OPHIUROIDEA
	Attached	PORIFERA BRYOZOA BRACHIOPODA BIVALVIA CRUSTACEA CRINOIDEA			COELENTERATA
Infaunal (living buried in the sea floor)	Burrowing in Soft Sediment	BIVALVIA ANNELIDA	BIVALVIA ANNELIDA OPHIUROIDEA ECHINOIDEA		GASTROPODA CRUSTACEA
	Boring in Hard Rock or Wood	PORIFERA BIVALVIA			

* Areas in gray indicate major groups of each feeding type.

some groups, such as the gastropods, bivalves, annelids, and crustaceans, have adapted to several habits, while others, particularly the attached forms such as corals, bryozoans, brachiopods, and crinoids, have only a single habit.

MARINE FOSSILS AND FOSSILIZATION

Having surveyed the life habits of all marine organisms, we shall now look more closely at those animals and plants that have hard, mineralized shells or skeletons, for these organisms are readily preserved in the fossil record and therefore make a disproportionately large contribution to our knowledge of the history of life in the sea.

Animals and plants use a variety of chemical compounds for the construction of shells and skeletons. One recent survey reports about a dozen, including such exotic ones as iron oxide (in the tooth structures of some molluscs), strontium sulfate, and barium sulfate (both in the skeletons of some Sarcodina). Most of these compounds are, however, either very rare or are found only in relatively insignificant groups of organisms; only three are of general occurrence as skeletal building materials: calcium carbonate ($CaCO_3$), silica (SiO_2), and calcium phosphate ($Ca_5(PO_4)_3OH$). The occurrence of these skeletal materials in the principal phyla of animals and plants is summarized in Table 3-3.

Skeletal Minerals

Calcium carbonate is by far the most abundant and widely distributed skeletal material, occurring commonly or frequently in 14 phyla of plants and invertebrate animals. Most of the shell-bearing animals of the sea make their shells of calcium carbonate, as do most mineral-depositing algae. Calcium carbonate is deposited by organisms in two crystallographic forms—*calcite* and *aragonite*. These two minerals have the same chemical composition, but differ in the geometric spacing of the atoms in the crystal; they are difficult to tell apart without an X-ray analysis. Because calcite is the more stable and less soluble of the two, calcitic fossil shells are usually better preserved than are those of aragonite. Although calcite occurs as a skeletal material in 13 phyla, aragonite is found in only 6 phyla; most of the phyla with aragonitic members also have members with calcitic shells. An individual shell is usually composed entirely of calcite or entirely of aragonite, but some snails and clams have shells that contain both minerals.

Silica is a much less widely distributed skeletal material than calcium carbonate. Two groups of small, unicellular organisms—the diatoms and radiolarians—build delicate lacy shells of silica, but the only larger organisms with siliceous skeletons are sponges, many of which are supported by a framework of needlelike silica rods. Except for a few large sponges, skeletons of silica are all very tiny; nevertheless they are common as fossils because they are hard and relatively insoluble.

Table 3-3 Distribution of Skeletal Minerals in the Principal Phyla of Organisms

Phyla		Calcium Carbonate CaCO$_3$		Silica SiO$_2$	Calcium Phosphate Ca$_5$(PO$_4$)$_3$OH
		Calcite	Aragonite		
Plants	SCHIZOMYCOPHYTA (bacteria) PYRROPHYTA (dinoflagellates)				
	CYANOPHYTA (blue-green algae)	frequent			
	CHLOROPHYTA (green algae)		frequent		
	CHAROPHYTA (stone worts)	frequent			
	PHAEOPHYTA (brown algae)				
	RHODOPHYTA (red algae)	common			
	CHRYSOPHYTA — diatoms			common	
	CHRYSOPHYTA — coccolithophorids	common			
	MYCOPHYTA (fungi) BRYOPHYTA (mosses) TRACHEOPHYTA (vascular plants)				
Animals	SARCODINA — radiolarians			common	
	SARCODINA — foraminiferans	common			
	PORIFERA (sponges)	frequent		common	
	COELENTERATA (corals)	common	common		
	BRYOZOA (bryozoans)	common	common		
	BRACHIOPODA (brachiopods)	common			frequent
	MOLLUSCA — snails	frequent	common		
	MOLLUSCA — clams	common	common		
	MOLLUSCA — cephalopods		common		
	ANNELIDA (segmented worms)	frequent	frequent		
	ARTHROPODA — trilobites crustaceans	common			common
	ARTHROPODA — arachnids insects				
	ECHINODERMATA (echinoderms)	common			
	CHORDATA — acorn worms tunicates, lancelets				
	CHORDATA — vertebrates				common

■ common   ▨ frequent   □ rare or absent

Although the final skeletal compound, calcium phosphate, occurs in only three groups, it is of extraordinary importance, for it makes up the bones and teeth of vertebrate animals. In addition, two invertebrate groups, brachiopods and arthropods, have some representatives with phosphatic skeletons. Unlike the vertebrates with *internal* bony skeletons, however, these invertebrates have *external* skeletons that do not resemble bone. Instead, the phosphate is interlayered with a tough organic material (similar to that of the human fingernail) that lends strength and flexibility to the skeleton. All phosphatic shells and skeletons are relatively stable and insoluble, and thus are commonly preserved as fossils.

You can see from Table 3-3 that silica occurs only in three primitive phyla (sponges and unicellular animals and plants) whereas calcium phosphate is found only in three advanced animal groups. Calcium carbonate skeletons, on the other hand, are found in some members of most of the phyla of animals and plants. The only major phyla that have few or no representatives with any kind of mineralized skeleton are all plant phyla: the Schizomycophyta (bacteria), Pyrrophyta (dinoflagellates), Phaeophyta (brown algae), Mycophyta (fungi), Bryophyta (mosses), and Tracheophyta (vascular plants). Even though these plant phyla lack mineralized tissues, many have tough organic coverings or woody skeletons of organic cellulose that may be preserved as fossils.

Preservation of Fossils

The original silica, calcium carbonate, or calcium phosphate of a shell or skeleton may be preserved without alteration in a fossil. More commonly, however, secondary changes occur in these materials after they are buried. Porous structures, particularly bone, may be filled with secondary silica, calcite, or other minerals deposited from waters in the surrounding sediment. Alternatively, the original mineral may *recrystallize*. Recrystallization is particularly characteristic of aragonite, which is unstable and commonly converts to calcite. Likewise, fine-grained original calcite is commonly altered to coarse-grained calcite by recrystallization. There can also be a complete *replacement* of the original mineral by a different secondary mineral. The most important replacement process is *silicification*, in which originally calcitic or aragonitic shells are completely replaced by secondary silica. Silicification is particularly common in fossiliferous limestones that have been exposed to the action of ground water. In such cases the silica selectively replaces the fossils but not the surrounding limestone matrix; thus it is possible to dissolve away the limestone in weak acids to leave only the enclosed siliceous fossils, which are not attacked by the acid. Acid treatment of silicified limestones sometimes yields large collections of exceptionally well-preserved marine invertebrates (Fig. 3-4). When recrystallization or replacement has occurred, it is sometimes difficult to determine the original shell mineral, but usually this can be predicted by comparing the specimens with better-preserved, unaltered fossils of the same kind. Finally, the original shell or skeletal

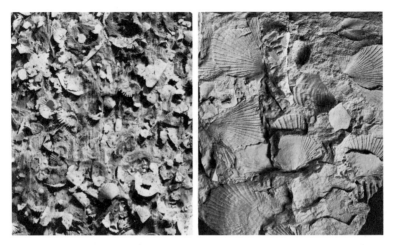

FIG. 3-4 (Left) Silicified Ordovician brachiopod shells, prepared by immersing a block of fossiliferous limestone in weak hydrochloric acid. Part of the limestone block has dissolved away, leaving the shells exposed on the surface. (Right) Natural molds of Devonian clam shells. The original shell material dissolved soon after burial, leaving only impressions in the surrounding sandstone. Both specimens are about four inches high.

material may be dissolved and not replaced, leaving only an impression or *mold* of the fossil in the surrounding rock (Fig. 3-4). Molds, which are particularly common in Paleozoic rocks, can often be as revealing as the original shell, for from them can be prepared artificial *casts*, made of rubber or plaster of Paris, that faithfully duplicate the original fossils.

All these modes of fossilization require that the original animal or plant had some hard, mineralized tissue to become fossilized; as we have seen, most phyla have many representatives with such tissues. We can get an overall review of the kinds of marine organisms that are most common as fossils by comparing the distribution of skeletal minerals shown in Table 3-3 with the information on modes of life in Table 3-1. The marine plant record is best for the planktonic forms; two of the three dominant groups of phytoplankton—the diatoms and coccolithophorids—have skeletons. The third major group of phytoplankton, the dinoflagellates, lack hard skeletons, but many have tough organic overings that are sometimes preserved. Benthonic algae have a relatively poor fossil record except for the red algae and some relatively small groups of green and blue-green algae. In general, the record of marine animals is much better than that of marine plants. All the dominant groups of marine invertebrates, both benthonic and planktonic, have skeletons except for some protozoans, all jellyfish, most annelids, and a few molluscs and crustaceans. Most marine vertebrates likewise have hard, phosphatic bones and teeth.

A hard shell or skeleton is the first essential for fossilization, but an equally important consideration for insuring the ultimate preservation of an organism is its life habit. Benthonic animals and plants, because they commonly live in areas of active sedimentation, are much more likely to be preserved by quick burial than are swimming and floating forms. In particular, the large, swimming vertebrates are rare as whole, articulated fossils because their bones and teeth are usually scattered by predators and scavengers after death. On the other hand, the shells of small planktonic organisms are relatively common fossils, for they tend to sink and accumulate on the bottom in great numbers after death. We can summarize, then, by noting that the fossil record of marine life is excellent for benthonic invertebrates, good for planktonic invertebrates and plants, and relatively poor for swimming animals (except shell-bearing cephalopods) and bethonic plants.

MARINE PLANTS

Having seen how organisms in the sea live and how they are preserved as fossils, we are now ready to return to the evolutionary history of marine life; for the remainder of this chapter we will consider the fossil record of marine plants and invertebrate animals. Although *marine vertebrates* have a long fossil record (fish have been common in the sea since the Devonian Period; marine reptiles and mammals appear after the close of the Paleozoic Era), they are of greatest interest for the light they shed on the origin and adaptive potential of land vertebrates and will therefore be considered in later chapters.

The geologic record of the seven principal phyla of marine plants is summarized in Fig. 3-5. You will recall from Chapter 1 that both major groups of procaryotes, the bacteria and the blue-green algae, have a long Precambrian record. Calcium-carbonate-depositing eucaryotic green and red algae have been reasonably common fossils since the Ordovician Period. The rarely fossilized eucaryotic brown algae are first known from a few doubtful impressions from Devonian rocks; the phylum may have originated earlier.

In striking contrast to the long fossil record of benthonic marine plants, the two primarily planktonic phyla are surprisingly late in first appearance. Fossil diatoms and coccolithophorids are first found in Jurassic rocks and do not become really abundant until Cretaceous time. The less commonly fossilized dinoflagellates may have had a similar history, for they first become abundant in Jurassic rocks. There is, however, a group of problematic microfossils (the "acritarchs"), found in rocks ranging in age from late Precambrian to Cretaceous, that may be distantly related to the dinoflagellates. The record of this group is shown as the broken extension of the dinoflagellate column in Fig. 3-5.

Dinoflagellates, diatoms, and coccolithophorids are today the dominant planktonic plants in the seas, where they account for the bulk of the photosynthetic production of nutrients that supply food for marine animal life. The

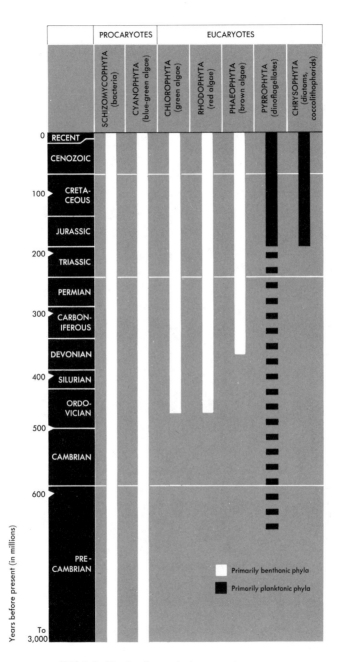

FIG. 3-5 The fossil record of marine plants.

late appearance of these fundamental groups is therefore puzzling, for marine *animals* similar to those found today were extremely abundant earlier, in Paleozoic time. We can only conclude that the primary sources of nutrients for these animals must have been different. This problem has so far received little attention; it seems most likely that planktonic plants were also abundant in Paleozoic time, but other than the problematic acritarchs, they did not secrete mineralized shells or tough cellulose coverings as do their dominant modern counterparts.

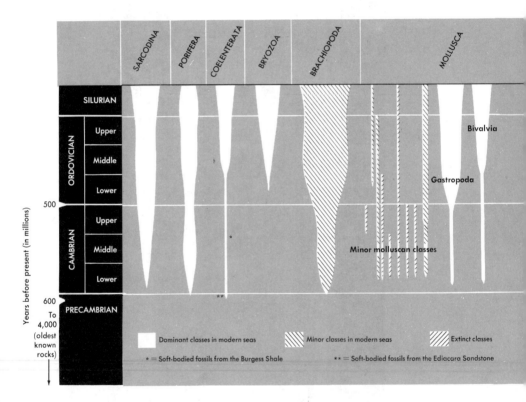

FIG. 3-6 Phyla and classes of Early Paleozoic marine invertebrates. The width of the white and striped areas indicates the approximate abundance of each group.

EARLY PALEOZOIC INVERTEBRATE LIFE

The geologic record of marine animals differs sharply from that of marine plants. The first appearances of the plant phyla are scattered through a tremendous span of time ranging from Early Precambrian to Jurassic. In contrast, *all but one phylum of invertebrate animals with preservable hard parts first appear at or near the Precambrian-Cambrian boundary* (Fig. 3-6). The exception is the Phylum Bryozoa, which first occurs near the middle of the Ordovician Period.

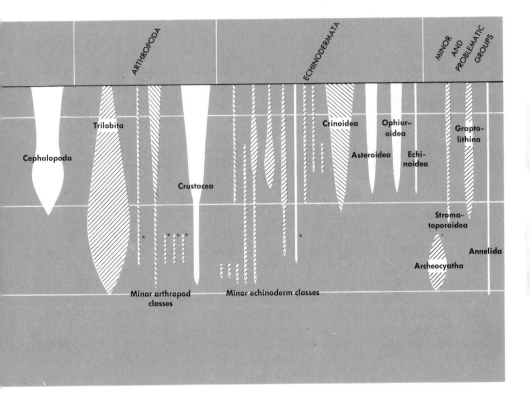

This is the reason for the dramatic contrast between the barren record of Precambrian animal life and the well-documented sequence beginning in Early Cambrian time. The major, phylum-level organizational patterns of invertebrate life were apparently determined in one evolutionary radiation during Late Precambrian and Early Cambrian time. Since then there has been extensive evolutionary change *within* the phyla, but, with the exception of the Bryozoa, no appearances of new invertebrate phyla.

The Principal Invertebrate Phyla

Eight phyla of invertebrate animals have many representatives with mineralized skeletons: Sarcodina, Porifera, Coelenterata, Bryozoa, Brachiopoda, Mollusca, Arthropoda, and Echinodermata. Fortunately for our understanding of invertebrate evolution, these eight phyla not only have an excellent fossil record, but also include the dominant invertebrates in the seas today. Only one soft-bodied phylum, the Annelida, is a significant contributor to modern marine invertebrate faunas. The seas have probably been dominated by the eight phyla with mineralized skeletons and the Annelida since Early Paleozoic time.

Unfortunately, there is no fossil record of the origin of these phyla, for they are already clearly separate and distinct when they first appear as fossils. Apparently, the common ancestors from which they arose either lacked preservable hard parts or were so rare and local in occurrence that they are not preserved in known Cambrian or Late Precambrian rocks. Because there are no intermediate fossils between phyla, speculations about their evolutionary relations must depend on indirect evidence, particularly on comparative studies of the anatomy, embryology, and biochemistry of present-day representatives. Such studies have led to conflicting interpretations, but there is reasonable agreement that the Bryozoa and Brachiopoda are more closely related to each other than to the other phyla. Likewise, the Mollusca, Arthropoda, and Annelida appear to be related, as do the Echinodermata and Chordata. The Sarcodina, Porifera, and Coelenterata do not appear to be closely related to any of the other phyla.

Origin of the Invertebrate Classes

If we now look at the next level of organization, the classes within the phyla, we see in Early Cambrian time an example of an evolutionary phenomenon that recurs often in later periods. Even though the modern *phyla* are clearly differentiated when they first appear as fossils, the modern *classes* are not. Instead, Early and Middle Cambrian rocks are filled with representatives of primitive classes that are now mostly extinct, having been replaced later by the more successful classes that dominate our modern oceans. The Early Paleozoic record of all fossilized invertebrate classes is shown in Fig. 3-6. Note that most of the modern shell-bearing classes first appear, along with the bryozoans, in Upper Cambrian and Lower Ordovician rocks. From Ordovician time to the present day, the seas have been dominated by such familiar groups as corals, bryozoans, snails, bivalves, cephalopods, and starfish. In contrast, if we could go back to explore a Cambrian sea we would mostly find certain arthropods called *trilobites*, which are extinct, and *brachiopods*, which still exist but are rare in modern oceans. It has been estimated that 60 percent of Cambrian fossils are trilobites and 30 percent brachiopods; all other groups make up the remaining 10 percent. Among this 10 percent are several classes of distinctive molluscs; many unusual echinoderms; several kinds of soft-bodied arthropods that resemble trilobites more closely than modern crustaceans; and a spongelike group called archeocyathids, which some paleontologists consider to be a separate, extinct phylum. The only Early and Middle Cambrian representatives of classes that are important today are some sarcodine protozoans, sponges, a few shrimplike crustaceans, rare clamlike and snaillike molluscs, soft-bodied jellyfish, and annelid worms.

We shall see that this is a common pattern in the history of the higher categories of life. Usually there is an early experimental stage in which many relatively short-lived and unsuccessful groups arise. The best adapted of these early groups then undergo a second radiation that leads to more successful and longer-lived groups. Within the major phyla of invertebrate animals, the Cambrian Period was the time of experimentation, and the Early Ordovician Period the time of secondary radiation and modernization. By Late Ordovician time most of the invertebrate classes that dominate the seas today were well established; since that time, changes in invertebrate life have been mostly smaller-scale evolutionary radiations and extinctions *within* these classes. Throughout the entire history of life, relatively few classes of organisms have ever become extinct. Most exceptions to this generalization are invertebrate classes that originated in Early Cambrian time, never became very common, and were extinct by the close of the Paleozoic Era.

A question of great evolutionary interest is the relationship of the various invertebrate classes to one another. Which primitive classes gave rise to the dominant modern ones? What were the transition forms like? We have seen that the phyla are already clearly differentiated when they first appear in the fossil record. A similar pattern exists within the classes. Unlike the intermediate forms between the phyla, which may have lacked hard parts, the ancestors of many of the modern shell-bearing classes were probably present among the more primitive shell-bearing classes of Early and Middle Cambrian time. Unfortunately, these extremely significant early fossils have only recently begun to receive critical evolutionary study. This work has already revealed several rare transitional forms between the classes, particularly among the molluscs and echinoderms, and we can expect the class-level interrelationships of many invertebrate groups to be clarified by future work on these neglected fossils.

Adaptations of Early Invertebrates

A closely related and equally neglected problem concerns the adaptations and life habits of the extinct Cambrian classes and the changes in habits that accompanied the rise of the modern classes. Similarities in shell form between Early Paleozoic and present-day representatives of the dominant modern classes make it reasonably certain that their adaptations have changed very little since they first arose. Sponges and bryozoans have probably *always* been principally epifaunal filter-feeders, cephalopods have been swimming carnivores, and so on. For this reason it is often possible to make detailed reconstructions of the life habits of fossil marine communities from Ordovician time to the present day (such reconstructions are discussed in another volume of this series, *Ancient Environments*). This situation changes sharply in the Cambrian record where so many classes lack modern representatives.

The key to understanding living communities of Cambrian invertebrates lies in understanding the habits of the trilobites that so strongly dominate the fossil assemblages. Unfortunately, very little attention has yet been directed toward reconstructing the habits of this important group. Their great diversity of shape and form suggests that they, like their modern relatives the crustaceans, had a variety of habits. Most were probably mobile herbivores or scavengers that moved around on the surface of the bottom. They were probably not active predators, for they lacked strong jaw structures, pincers, or other predatory adaptations. Some may have been sediment-feeders and a few may even have been filter-feeders. If, as seems likely, the trilobites were mostly herbivores or scavengers, their great abundance suggests that there must have been a variety of unpreserved benthonic algae and soft-bodied invertebrates to supply their food. Cambrian brachiopods and sponges, like their modern relatives, were almost certainly epifaunal filter-feeders. The probable life habits of the other Cambrian groups is a promising but as yet unexplored subject.

The Burgess Shale Fauna

Although Cambrian fossil assemblages are dominated by shell-bearing trilobites and brachiopods, we know that soft-bodied invertebrates were also common because of two remarkable occurrences of well-preserved, soft-bodied fossils in pre-Ordovician rocks. The older of these, the Late Precambrian Ediacara fauna of Australia, was discussed in Chapter 1. Recall that the fauna contained impressions of soft-bodied coelenterates and annelid worms.

The other occurrence of early soft-bodied fossils was discovered in 1910 by C. D. Walcott, director of the U.S. Geological Survey and a pioneer student of Cambrian paleontology and stratigraphy. While crossing a steep mountain pass near Field, British Columbia, Walcott and his party stumbled on some slabs of dark shale containing both Middle Cambrian trilobites and clear impressions of soft-bodied, wormlike fossils. Investigation revealed that the fossils occurred in several poorly exposed shale layers on the steep slope of the pass. In the following summers Walcott returned to the locality with a large field party that blasted and quarried 22 feet into solid rock to collect the fossils, most of which were found in a single three-foot layer (Fig. 3-7). Preliminary studies by Walcott revealed that the fauna contained over 100 species of well-preserved, soft-bodied invertebrates in addition to many shell-bearing trilobites and sponges. Included were: four primitive classes of soft-bodied but generally trilobitelike arthropods; some jellyfish and anemonelike coelenterates; 11 genera of probable annelid worms; and a possible early holothurian, or "sea cucumber," the only present-day echinoderm class that lacks a skeleton. Walcott named the fossil-bearing formation the "Burgess Pass Shale," and the occurrence has come to be

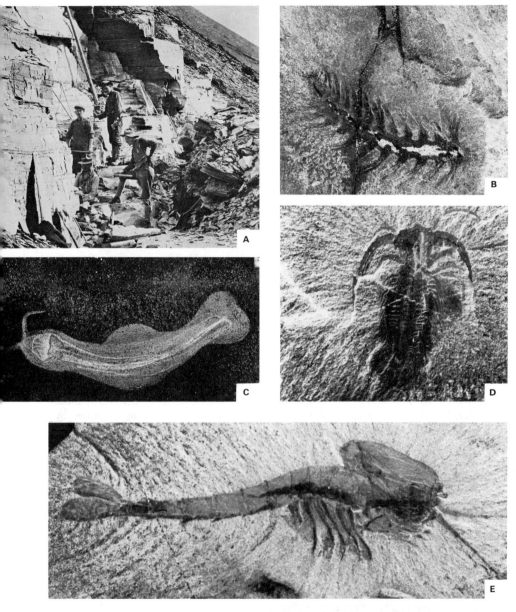

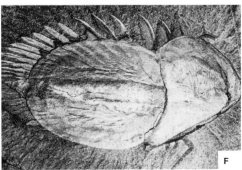

FIG. 3-7 (A) Walcott's quarry in the Cambrian Burgess Shale near Field, British Columbia. Discovered in 1910, the quarry yielded about 35,000 specimens of unique soft-bodied invertebrates. (B–C) Annelid worms and (D–F) soft-bodied arthropods from the quarry. (Courtesy Smithsonian Institution.)

known as the Burgess Shale fauna. Like the earlier Ediacara fauna, the Burgess Shale fauna contains soft-bodied coelenterates and annelids. Unlike the Ediacara, however, it is dominated by arthropods, both trilobites and unique representatives of four primitive nonshelled classes. Walcott's expeditions collected over 35,000 specimens of these extremely important fossils, but, surprisingly, they have received very little critical study since Walcott's day. Fortunately, a program of modern recollection and revision of the fauna has been initiated by the Geological Survey of Canada, and, as a result, we can expect to learn much more about these unique and extremely significant fossils.

The Ordovician Expansion of Shell-Bearing Invertebrates

Both the Ediacara and Burgess Shale faunas show that soft-bodied invertebrate life was abundant in pre-Ordovician seas, even though shell-bearing brachiopods and trilobites dominate the fossil record. This fact raises the question of the origin and function of shells in these early invertebrates. The problem is really twofold: Why did shells develop in the first place in Early Cambrian time, and why did shell-bearing groups rapidly expand and diversify about 100 million years later in Early Ordovician time?

As we saw in Chapter 1, there is no completely satisfactory answer to the first question, although many theories have been proposed. In contrast, very little attention has been given to the related question of the Ordovician expansion of the modern shell-bearing groups. It has been suggested that chemical changes in the Early Ordovician seas might have favored deposition of mineralized skeletons, but the abundance of shell-bearing Cambrian trilobites and brachiopods indicates that earlier sea water was also chemically favorable for shell growth. A more plausible suggestion involves the role of shells as protective structures. Shells and skeletons have two principal functions in modern marine invertebrates: support for the soft tissues of the animal and protection against predators. In many groups the functions are combined (bivalves, brachiopods); in others the shell is dominantly either protective (gastropods) or structural (most crustaceans). We can postulate that shells arose in Early Cambrian time primarily as a means of structural support, and that, although the experiment was relatively unsuccessful in most groups (including the rare early echinoderms and molluscs), it led to expansion in trilobites and brachiopods. Along with these early groups with supporting skeletons, there also arose a host of nonshelled arthropods, coelenterates, annelids, and other wormlike groups. Indeed, these soft-bodied forms may have played a more important role in Cambrian seas than at any time since the Ordovician expansion of the dominant shell-bearing groups. Predators were probably rare. Among known fossils of Early and Middle Cambrian time, none seem to have had adaptations for active predation. In fact, the first predatory fossil cephalopods do not appear until Late Cambrian time; predatory fish appear much later, in Upper Silurian rocks. It may be significant that predatory cephalopods arose near the time of rapid expansion of the modern shell-bearing classes. The appearance of these active and efficient carnivores may

Cretaceous Period. Particularly affected were the shelled cephalopods, which had an erratic evolutionary history throughout the Paleozoic and Mesozoic Eras, during which repeated extinctions of entire subgroups were followed by rapid evolutionary expansion of new groups. Shelled cephalopods are represented today by only a single genus, *Nautilus*; on the other hand, unshelled cephalopods (squids and octopuses) have expanded to dominance during the Cenozoic Era.

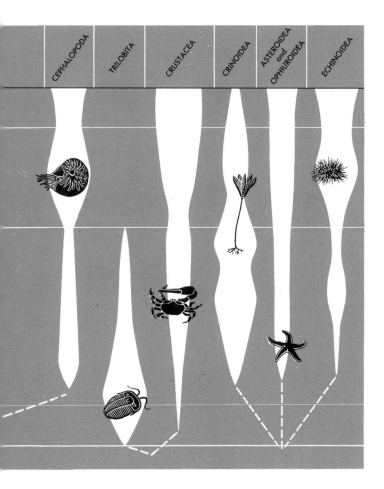

POSSIBLE CAUSES FOR WIDESPREAD
EXTINCTIONS

The pattern of extinctions and subsequent evolutionary radiations of marine invertebrates is seen most dramatically at the boundaries separating the three eras, yet less pronounced intervals of widespread extinction mark the boundaries between most of the major units of geologic time. This raises the question of the causes of extinctions and evolutionary radiations. The evolutionary bursts that follow periods of extinction can be readily explained, for it seems reasonable that the modes of life and environments vacated by the extinct organisms would provide fertile ground for new evolutionary experimentation and expansion. The more basic problem appears to be the cause of the extinctions. The problem is not just confined to invertebrate life, for vertebrates and plants tend to be affected by the same periods of crisis as marine invertebrates. Whatever the causes, they must be broad enough to affect the entire living world.

These patterns have led to a host of explanations, but none has been generally accepted. Certainly, the answer must lie in changes in the environment and in the failure of organisms to adapt to these changes, but the nature of the changes is obscure. Among the more probable hypotheses that have been proposed to account for periodic extinctions are: changes in the concentration of atmospheric oxygen, excessive radiation from space, development of pathogenic organisms, excesses or deficiencies of trace elements in the sea and soil, and changes in climate, the latter being the most frequently suggested cause. Climatic changes have certainly occurred and have undoubtedly affected organisms, yet they seem to offer an incomplete general explanation for extinctions because the most recent several million years of Earth history have been marked by unusually severe climate fluctuations, yet these have *not* been accompanied by wholesale animal and plant extinctions. This fundamental and poorly understood problem will be discussed again in later chapters as we follow the origin and development of life on land.

four

the transition to land

Life originated in the sea, and most phyla, particularly those of algae and invertebrate animals, are still predominantly or exclusively marine. It was not long after the great proliferation of marine life in Early Paleozoic time, however, that the first plants and animals began to colonize the lifeless surface of the land, for the oldest land fossils are found in Silurian rocks.

The problems faced by early sea-dwelling organisms in making the transition to land were many and formidable. Whereas animals and plants living in the sea have an inexhaustible supply of water, land-dwellers must obtain water from rain, streams, soil, or the food they eat. In addition, to prevent the evaporation of the water they get, they must have tough, relatively watertight coverings, such as the waxy surface of leaves or the skin of reptiles and mammals. Land animals also require special structures for breathing the oxygen of the air rather than absorbing oxygen from the surrounding water. Reproduction presented still other types of problems. In the sea, organisms normally shed their gametes directly into the water for fertilization. On land, special adaptations are required to prevent the drying out of the delicate gametes. In spite of these difficulties, the advantages of colonizing the land were considerable. For plants, there was abundant unoccupied space with direct sunlight for photosynthesis. For animals, there was the abundant free oxygen of the atmosphere and, after the spread of land plants, an almost limitless supply of food.

Most land-dwelling organisms probably made the transition from the sea by way of the fresh waters of rivers and lakes—that is, they first adapted to water lacking the dissolved salts of the sea, and only then developed further adaptations for life out of water. The transition from salt to fresh water did not present the hazards involved in obtaining water and reproducing, but for animals it involved an almost equally serious problem—that of maintaining the balance of salts in the body fluids. The blood and other body fluids of all animals contain about the same amounts and kinds of dissolved salts as does sea water, an amazing fact in itself suggesting that all animal life originated in the sea. Marine animals, because of this similarity, have no difficulty in maintaining the proper balance of body salts, whereas freshwater and land-dwelling animals (which, of course, also take in mostly fresh water) must develop special organs and expend a great deal of energy to prevent a fatal dilution of the body fluids. For this reason, even the transition from the sea to rivers and lakes presented a serious adaptive problem.

As you might expect, more phyla have overcome the problem of salt balance in fresh water than have developed the additional adaptations needed for life in the open air. Many of the major phyla—including the Schizomycophyta, Chlorophyta, Charophyta, Chrysophyta, Sarcodina, Mollusca, Annelida, Arthropoda, and Chordata—have representatives adapted to fresh water. Only four phyla, however, have members that are fully adapted to life on dry land: Tracheophyta (all), Mollusca (some snails), Arthropoda (arachnids, insects), and Chordata (reptiles, birds, mammals). In this chapter we will survey the transitional stages in the adaptation of these four phyla to life on land before discussing, in the remaining chapters, the evolutionary history of land-dwelling plants and vertebrate animals.

PLANT TRANSITIONS

Botanists generally agree that green algae of the Phylum Chlorophyta are the most probable ancestors for the Phylum Tracheophyta, which includes all the dominant land-dwelling plants, such as ferns, conifers, and flowering plants. The two phyla show close similarities in their manner of photosynthesis: both have the same green pigments, and both yield the same carbohydrate as the photosynthetic end product. No other algal phylum shares these biochemical characteristics with the tracheophytes. In addition, the green algae are one of the few algal groups to have made the transition successfully from marine to fresh waters, the usual first step toward life on land. Today, green algae are extremely abundant in fresh water (the familiar green "pond scum" is an example), and some are even adapted to life in moist soils or on damp tree trunks.

Unfortunately, there is no fossil record of the transition between an algal ancestor and the first land-dwelling tracheophytes. The transitional forms probably lacked the calcified structures of some of their marine algal relatives as well as the tough cellulose stems of their later land-dwelling descendants. There are, however, some familiar modern plants that show instructive transitional characteristics, even though they are probably not directly related to the ancestral land plants. These are the mosses and liverworts of the Phylum Bryophyta, a group of truly "amphibious" plants with ties both to land and to water. Bryophytes live primarily in moist, shady environments because they lack two features required for fully successful land life: (1) an efficient system for transporting fluids within the body, and (2) a means of preventing the gametes from drying out.

Some kind of fluid-transport system is essential for all land plants. Aquatic plants, both freshwater and marine, can absorb water and mineral nutrients directly over the entire body surface, but land-dwelling plants must obtain them from the soil. Photosynthesis, however, cannot take place beneath the soil; it requires sunlight. As a result, land plants have developed specialized organs for each function: *roots* to gather water and nutrients from the soil, and chlorophyll-bearing *leaves* for photosynthesis above the ground. In addition, to grow very large, a plant requires a system for transporting fluids between the roots and leaves, and it also must have a means of preventing loss of fluids through the surface. All land plants of the Phylum Tracheophyta are characterized by a *vascular system* composed of special fluid-conducting cells in the stem. Along with the vascular system, there developed a special surface layer of waxy material, called the *cuticle*, that prevents the loss of water from the surface of the stems and leaves.

Mosses (Fig. 4-1) and liverworts are unique in having developed both rootlike and leaflike structures without a vascular system to transport fluids between these organs. They are all very small organisms, for they must depend on simple diffusion for fluid transport. The largest reach a height of only about a foot and most are only a few inches tall. In addition, because the mosses and liverworts have not developed seeds or other specialized structures to protect the gametes from drying out, they require a moist environment for reproduction.

Rare fossil liverworts are found in Devonian rocks, but these are apparently not in the direct line of evolution of vascular plants, traces of which occur still earlier. There have been several reports of poorly preserved spores and stems

FIG. 4-1 Mosses, "amphibious" land plants that live mostly in moist environments. These plants never grow very large because they lack an efficient system for internal transport of water and nutrients. (Courtesy American Museum of Natural History.)

of presumably vascular plants from Cambrian and Ordovician rocks, but these Early Paleozoic occurrences are all questionable; the first undoubted vascular plants are found in Upper Silurian rocks. These earliest land plants were still very simple forms, but they had a rudimentary vascular system that permitted growth upward into the open air and thus paved the way for a rapid evolutionary expansion; by mid-Devonian time much of the land surface was covered by great forests of vascular plants.

TERRESTRIAL INVERTEBRATES

Many phyla of invertebrate animals have become adapted to life in rivers and lakes. Particularly successful are protozoans, clams, snails, crustaceans, and several phyla of soft-bodied, wormlike invertebrates, some of which, like the earthworm, are also found in moist soils. There are even a few freshwater sponges, coelenterates, and bryozoans, but many important invertebrate groups —for example, brachiopods, cephalopods, and echinoderms—have never left the sea. Although invertebrate adaptation to fresh water has occurred in many phyla, only two, the molluscs and arthropods, have representatives that are fully adapted to life out of water. Even among the molluscs, only gastropods have made the transition to become "land snails." The arthropods have been by far the most successful land invertebrates. Of these, some fully terrestrial crustaceans have evolved, but it is two other arthropod groups—the arachnids and particularly the insects—that are the invertebrate masters of the land.

The success of both gastropods and arthropods in invading the land is largely due to adaptations already present in their aquatic ancestors. Gastropods had a watertight shell into which they could withdraw for protection against drying out. Likewise, arthropods had a tough, waterproof, organic covering. In addition to having these "preadaptations" against water loss, aquatic gastropods and arthropods are mostly mobile animals and thus were adapted to moving about in search of food on land.

The problem of respiration in the open air was solved differently by the two groups. The gastropods developed lungs, comparable to those found in land-dwelling vertebrates, for oxygenating the blood. Insects and arachnids, in contrast, developed a series of tubelike structures called *tracheae* that open to the air as tiny holes in the organic covering; these tubes conduct oxygen directly to the internal body tissues. These animals thus breathe over much of the body surface and do not require a complex and cumbersome circulatory-respiratory system such as that of the air-breathing snails and vertebrates. At the same time, however, breathing by means of tracheae imposes limits on maximum size, for the direct diffusion of oxygen rapidly becomes ineffective as size increases. This is the reason that insects and arachnids tend to be relatively small animals.

Both gastropods and arachnids remained "benthonic" on land, for both move and feed primarily on the surface of the ground or on plants growing in

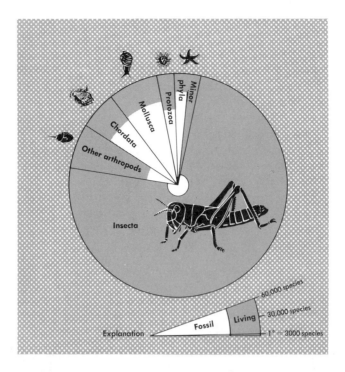

FIG. 4-2 The relative abundance of living and fossil insects. Insects make up about three-fourths of all living animal species, but only a tiny fraction of fossil species. The relative proportions of fossil and living species for each group are shown by the white (fossil) and gray (living) areas of the circle. (Modified from Moore, 1955.)

the ground. Insects, however, obtained freedom from the land surface by developing wings for flying in the fluid air above the ground, an adaptation that is in many ways comparable to the development of swimming in the more viscous water of the sea. This ability to move rapidly in the air to search for food and flee from predators was a tremendous adaptive advantage. It led, indeed, to the most phenomenal evolutionary radiation of all, for insects include about three-fourths of all living animal species (Fig. 4-2). Although they represent only a single class of one phylum, insects are the most successful organisms ever to evolve, if we measure evolutionary success by numbers of species and individual animals.

In contrast to the excellent fossil record of marine invertebrates, there are relatively few fossils of freshwater and land-dwelling forms, partly because two of the dominant groups, the insects and arachnids, lack mineralized skeletons. In addition, few terrestrial invertebrates live in areas of sediment accumulation where they can be readily buried and preserved. Of those that do, freshwater crustaceans and clams are reasonably common fossils in ancient lake and river deposits since the Devonian Period; air-breathing gastropods, although they are first found in Carboniferous rocks, are common fossils only in some nonmarine Cenozoic rocks.

Because of the great importance of insects in our modern world, their evolutionary history is of unusual interest. Even though they are rarely fossilized, a few dozen localities around the world have yielded large faunas of insects that help us understand their history. These fossils are preserved in two different ways. Most are delicate impressions, usually of the wings, preserved in fine muds that accumulated in ancient lakes or swamps. The fragile animals were apparently protected from decay by falling into the water and being rapidly buried in the bottom mud. Still more dramatic are insects and spiders preserved in amber, the fossilized resin of ancient trees. Animals preserved in amber sometimes retain even the finest details of soft structure, such as muscles and hairlike bristles (Fig. 4-3). Some even still have parasitic internal worms! Most such insects are known from amber that washes up on the shores of the Baltic Sea from submerged rocks, thought to be of Miocene age; occasional insect-bearing amber deposits are found in Tertiary and Cretaceous rocks of other regions.

FIG. 4-3 Fossil insects preserved in amber. (Courtesy Field Museum of Natural History.)

The oldest-known land animals are rare Early Devonian arachnids and, less certainly, insects, but both groups apparently first became really abundant in the Carboniferous Period, where large faunas are sometimes found in shales associated with coal deposits. Some of these early insects were primitive flightless forms that lacked wings, yet the ability to fly must have developed very early in insect history, for Carboniferous and Permian rocks already contain many flying forms, including a dragonflylike species with a three-foot wingspread that is the largest insect known. Unfortunately, the earliest fossil record of insects, arachnids, and terrestrial gastropods is too spotty to document the transition from their water-dwelling ancestors; instead, these rarely preserved groups are already fully developed terrestrial forms when they first appear in the fossil record. Only when we move on to vertebrate animals is the transition to land clearly recorded.

ORIGIN OF THE VERTEBRATES

The history of invertebrate animals is of great evolutionary importance because invertebrates far outnumber vertebrates in terms of phyla, classes, species, and individuals, both today and in the past. Nevertheless, it is the history of vertebrate animals that is of greatest interest to us, not only because we ourselves are vertebrates, but also because fishes, reptiles, birds, and mammals are much more familiar—as food, pets, and sources of recreation.

Vertebrates certainly arose from some sort of invertebrate ancestor, but, as is so often the case, the exact ancestral group is uncertain. The fossil record provides no clues because the earliest fossil vertebrates, bone fragments of primitive fishes found in Middle Ordovician rocks, are already fully differentiated from their invertebrate ancestors. Our knowledge of the origin of water-dwelling vertebrates must therefore depend entirely on indirect evidence, the most instructive of which is provided by some present-day soft-bodied marine animals that appear to be related to the ancestral vertebrates. These animals have a backbonelike structure at some stage of their life history, but, unlike that of the vertebrates, it is not divided into separate segments, or "vertebrae." Instead it is a solid rod, called a "notochord," made up of stiff, gelatinous organic matter. A similar solid notochord is found in the early embryological development of the vertebrates, but it is replaced by separate vertebrae in later stages. These notochord-bearing animals have almost no fossil record, although three different kinds are found in modern seas. The three groups, along with the vertebrates, make up the four subphyla of the Phylum Chordata, animals with backbones. The three notochord-bearing subphyla can be thought of as "invertebrate chordates," since they have a backbonelike structure but lack separate vertebrae.

Invertebrate Chordates

The most revealing of the invertebrate chordates are the *lancelets*, elongate, almost transparent animals that make up the Subphylum Cephalochordata (Fig. 4-4(A)). Lancelets are small animals (the largest are only three or four inches long) that live partially buried in the sand along the shores of warm seas. In addition to a well-developed notochord, lancelets have many gill slits behind the mouth; along with their elongate shape and ability to swim when not buried, these slits make them very fishlike in appearance. In most aspects of their anatomy, however, they are much simpler than even the simplest vertebrate. Lacking jaws and teeth, they feed by extracting fine particles from the water as it passes through the gills. They have simple nervous and digestive systems and lack such features as paired fins, eyes, and a differentiated head and brain, all of which are characteristic of fishes. Lancelets probably are somewhat specialized

A

B

C

FIG. 4-4 Invertebrate chordates, simple relatives of the vertebrates with a solid backbonelike structure that is not divided into separate vertebrae. (A) Amphioxus, a fishlike lancelet of the Subphylum Cephalochordata. (Courtesy American Museum of Natural History.) (B) Spongelike tunicate of the Subphylum Urochordata. (C) Acorn worms of the Subphylum Hemichordata. (Both courtesy Ralph Buchsbaum.) (D) Life habits of invertebrate chordates. Acorn worms (left), and lancelets (center), mostly live partly buried in sediments; most tunicates (right), attach to hard objects (some are also planktonic floaters). All three groups are suspension-feeders found only in the sea. The specimens shown are about three inches long.

D

descendants of the Early Paleozoic ancestors of the vertebrates, and thus give us a revealing glance at the prefish stage of vertebrate evolution.

The other two subphyla of invertebrate chordates are also probably descended from the ancestral vertebrates of Early Paleozoic time, but they have become more specialized than the lancelets and lack their suggestive, fishlike characteristics. One group, the Subphylum Urochordata, is made up of attached or floating marine suspension-feeders that superficially resemble sponges (Fig. 4-4(B)). A notochord is present in the larvae, but not in the adult. The other group, the Subphylum Hemichordata, or "acorn worms," are wormlike animals (Fig. 4-4(C)) that live in tubes they construct in mud and sand on the sea floor. Like the lancelets, they have many gill slits that serve to filter small particles from the sea water. The adults appear to have a rudimentary notochord, but it is the early larval stages that are of greatest interest, for they give a clue to a still earlier step in vertebrate history—the relationship of these invertebrate chordates to the other invertebrates.

Scientists have speculated for a long time about the most likely invertebrate ancestor of the chordates. Perhaps the favorite candidates have been the annelids and arthropods, because their segmented bodies vaguely suggest the many vertebrae and bones of the vertebrate skeleton. There are serious difficulties, however, in deriving the vertebrates from either of these groups, which are now considered unlikely ancestors. Instead, it appears that the chordates may be most closely related to the echinoderms. Typical echinoderms, such as starfish and sea urchins, seem like most improbable relatives of the vertebrates, yet the larval stages of echinoderms are so similar to those of the notochord-bearing acorn worms that for many years young acorn worms were identified as starfish larvae. More important still, the larval stages of acorn worms and echinoderms, while closely similar to each other, are quite unlike the larvae of all other invertebrates. In addition to these larval similarities, there are biochemical and embryological parallels that also suggest a relationship between the echinoderms and chordates. It is unlikely, however, that chordates could have evolved from any known group of fossil or living echinoderm. Instead, it seems more probable that both groups evolved from an as yet unrecognized ancestor in earliest Paleozoic time.

The notochord of the ancestral vertebrates and the internal skeleton that developed from it were key adaptations that paved the way for the extraordinary evolutionary success of the vertebrates. In almost all invertebrate animals the skeleton is external, a condition that places great limitations on the strength of the skeleton and the ultimate size and mobility of the animal. In contrast, the internal skeleton of the vertebrates served as a strengthening support system, yet could remain relatively light and flexible because it need not cover the entire exterior of the body. The internal skeleton, and the efficient muscle and nervous systems that developed with it, permitted the development of many large and mobile vertebrate animals that are without parallel in the invertebrate world.

THE EVOLUTIONARY HISTORY OF FISHES

Although there is no fossil evidence for the origin of the vertebrates, the history of the group, once it was differentiated, is documented by an unusually complete record. The earliest vertebrates, from which all others arose, were primitive fishes. Indeed, four of the eight classes of vertebrates are fishes, and today, as in the past, fishes far outnumber land-dwellers in numbers of both species and individuals. Most of the evolutionary history of fishes is, however, a side issue in the mainstream of vertebrate development that leads to land-dwelling amphibians, reptiles, birds, and mammals. For the remainder of this chapter we will consider the history of fishes and of amphibians, those transitional land-dwelling vertebrates that arose from fishes but still had close ties to the water. Chapters 6 and 7 will continue our chronology of vertebrate development by discussing the three most successful classes of land-dwelling vertebrates —the reptiles, birds, and mammals.

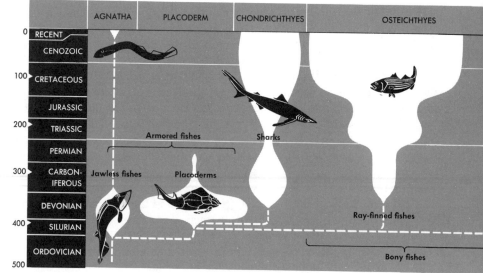

Although the first fragments of fossil fish bone are found in Middle Ordovician rocks, remains of fishes are rare and poorly preserved until the end of the Silurian Period, when fishes began an explosive evolutionary radiation that lasted throughout the succeeding Devonian Period (Fig. 4-5). Just as with the invertebrates of the preceding Cambrian and Ordovician Periods, fishes first went through a stage of evolutionary experimentation that, for the fishes, lasted from Late Silurian to Middle Devonian time. This phase of fish evolution was dominated by two primitive and now mostly extinct classes, Agnatha and Placodermi.

From these early experimental groups arose the two dominant classes of modern fishes, the Chondrichthyes (sharks) and Osteichthyes (bony fishes), which rapidly replaced the earlier classes in Late Devonian time and have been dominant ever since. Note that most of the evolutionary expansion and replacement of the four fish classes took place during the Devonian Period, a key time in fish evolution. For this reason the Devonian Period is sometimes called the "Age of Fishes," even though fossil fishes are much less common than invertebrate animals in Devonian rocks.

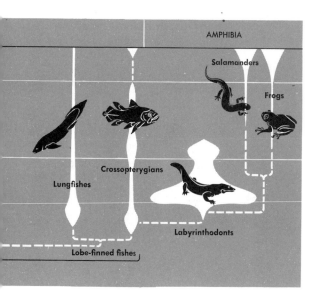

FIG. 4-5 Evolutionary history of fishes and amphibians. (Modified from Schaeffer in Millot, 1955.) The dashed lines show the most probably evolutionary relations of the group. The width of the white areas indicates the approximate abundance of each group.

Agnaths

The evolutionary histories of the four classes of fishes are summarized in Fig. 4-5. The first and most primitive of the classes were the agnaths, or "jawless fishes" (Fig. 4-6), a group that includes all but a very few of the Ordovician and Silurian fossil fishes. Like their probable lanceletlike ancestors, these first vertebrates had not yet developed a key vertebrate adaptation—jaws for seizing and chewing food. Without jaws, these early fishes could not feed on other animals or even on aquatic plants; instead they mimicked their invertebrate ancestors by remaining deposit- or suspension-feeders. There are several different types of these early fishes, but, from the shape of their skulls and skeletons, it is evident that most were deposit-feeders that moved slowly along the bottom shoveling organic-rich sediments into their jawless mouths. Some, however, appear to have been suspension-feeders with adaptations for straining plankton from the surface waters. Whichever their feeding style, they were mostly small animals, less than a foot in length, and nearly all were covered by a tough coat of bony plates. Indeed, were it not for these external plates, agnaths would now be unknown as fossils; the internal skeleton is never preserved and was apparently composed of uncalcified organic cartilage. This condition is not uncommon in other groups of fishes and is characteristic of all members of the Class Chondrichthyes (sharks and their relatives). The function of the thick bony armor of these early fishes is uncertain. In salt water, the armor might have served as protection from predation by cephalopods or arthropods. Armor was

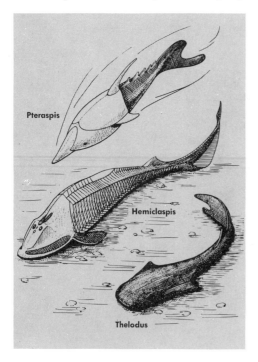

FIG. 4-6 Silurian and Devonian armored jawless fishes, of the Class Agnatha, were deposit-feeders that ingested organic-rich bottom sediments with their shovellike mouths; those shown were about a foot long.

Pteraspis

Hemiclaspis

Thelodus

also present, however, in early freshwater forms that lived where there were no known predators. It has been suggested that the armor of these freshwater fishes helped to keep them from losing essential body salts to the surrounding salt-free water; but a tough, unarmored skin would have served the same function and would seem to have been a less complex and more readily evolved adaptation. Whatever the function, heavy external armor was characteristic not only of the early agnaths but also of the placoderms, the first jaw-bearing fishes that evolved from them.

The jawless agnaths were the only vertebrates throughout the Ordovician and most of the Silurian Periods, but they were never very common. Apparently, deposit-feeding kept them a rather small and insignificant group. Although they persisted throughout the Devonian Period, they declined rapidly after the expansion of their more successful, jaw-bearing offspring in Early Devonian time. All the armored agnaths were extinct by the close of the Devonian Period, yet, strangely enough, the agnathan group survives today because some early unarmored representatives adapted the jawless mouth to a new mode of feeding—parasitism. Living lampreys and hagfishes, which attach to and suck the blood of other vertebrates with jawless mouths, are thought to be specialized descendants of the early agnaths. These modern parasitic agnaths have neither armor nor calcified skeletal parts and are unknown as fossils.

Placoderms

Jaws were an adaptation second only to the internal skeleton in importance to the success of the vertebrates, for they permitted the development of carnivorous and, ultimately, herbivorous feeding and thus paved the way for the rise of land vertebrates. Although no transitional fossils are known, placoderms, the first jaw-bearing fishes, almost certainly originated from the agnaths. Comparative studies of fossil agnath and placoderm skulls, combined with observations on jaw development in the embryos of living vertebrates, show that placoderm jaws probably evolved from the bony supports of the anterior gill of the jawless agnaths.

Placoderms, first found in Late Silurian rocks, went through a rapid radiation in Early Devonian time. By the middle of the Devonian Period they were extremely diversified. Most dramatic were the *arthrodires*, a group of carnivorous marine placoderms that includes one of the most fearsome predators of all time, a huge-jawed, thirty-foot monster that must have been the scourge of Devonian seas as it fed on its smaller placoderm relatives (Fig. 4-7). Most placoderms, however, were smaller forms, from one to a few feet in length, that fed on bottom-living algae and invertebrates, and on each other. Like their agnath ancestors, most placoderms had a tough, bony armor whose exact function is unknown. Unlike the agnaths, however, placoderms had partially calcified vertebrae and internal bones that are often fossilized. The evolutionary dominance of the placoderms was brief. They arose in Late Silurian time, reached

FIG. 4-7 A giant Devonian arthrodire, jawed carnivorous fish of the Class Placodermi. *Dinichthys,* the genus shown, reached a length of 30 feet and had a mouth several feet wide. It probably ate smaller placoderms.

their peak of abundance and diversity in the Middle Devonian, and began their decline in Late Devonian time as they were replaced by their dominant modern descendants, the sharks and bony fishes of the classes Chondrichthyes and Osteichthyes. A few placoderms persisted until the Permian Period, but the group is unknown after the close of the Paleozoic Era; it is, indeed, the only extinct class of vertebrates.

Before looking at the history of sharks and bony fishes, the dominant aquatic vertebrates today, let us consider whether the earliest agnaths and placoderms originated in fresh or salt waters. We have already seen that all the principal invertebrate phyla originated in the sea. Modern invertebrate chordates—the lancelets, tunicates, and acorn worms—are also found only in the sea, which suggests that the vertebrates also may have had a marine origin. Many Devonian fossil fish, however, are found in rocks that were apparently deposited in lakes and rivers on the land, suggesting, in contrast, that vertebrates might have originated in fresh water and only secondarily returned to the sea. But it now appears that most fossil agnaths and placoderms from Ordovician and Silurian rocks are associated with marine invertebrate fossils, whereas only in the Devonian Period do fishes become common in nonmarine rocks. Thus, a marine origin for the vertebrates appears most likely. There is no question, however, that by Early Devonian time both agnaths and placoderms had become adapted to fresh waters, and, indeed, the Devonian evolutionary radiation of fishes probably took place largely in rivers and lakes on land. Both marine and fresh waters, then, have been populated with fishes since the Devonian Period.

Sharks and Bony Fishes

The sharks and bony fishes of the Classes Chondricthyes and Osteichthyes differ from the placoderms in two major respects: they developed a more efficient jaw mechanism, and they became much more effective swimmers by losing the heavy external armor and by developing more efficient fins and body muscles. Both groups first appear in the Devonian Period and both probably arose from the placoderms, although, once again, intermediate fossils are lacking. The two groups also differ fundamentally from each other. Sharks lack a calcified skeleton; instead, the vertebrae and internal bones are made entirely of tough, organic cartilage. For this reason the class has a relatively poor fossil record. The teeth and some external spines of sharks are, however, calcified, and "sharks' teeth" are common fossils in many marine sedimentary rocks. In contrast, the bony fishes, as the name implies, have a strongly calcified bony skeleton that is frequently fossilized. There is one other fundamental difference between the two groups: sharks apparently arose from marine placoderms and have remained in the sea, whereas bony fishes appear to have developed from freshwater placoderms and to have lived predominantly in fresh waters throughout their early history. Both sharks and bony fishes have expanded steadily from their origin in the Devonian to their present dominance. Sharks have mostly remained aggressive, carnivorous predators, but the bony fishes, in addition to predatory forms, have developed many scavenging, herbivorous, and bottom-dwelling groups. Many of these varied forms arose during the Cretaceous Period in a second great evolutionary radiation that led several groups of Mesozoic bony fishes to reinvade the sea where they competed successfully with the sharks. Today marine species of bony fishes far outnumber freshwater species.

When the bony fishes first appeared in the Devonian Period they were already differentiated into two groups that were to have divergent evolutionary roles. Between these two groups there were disparities in the structure of the skull and skeleton, but the most important distinction was in the fins on the

FIG. 4-8 Arrangements of fin bones in typical lobe-finned (left) and ray-finned (right) fishes. In the lobe-fins, muscles extended into the fin, permitting greater control and flexibility of movement. Such fins developed into the limbs of the land-dwelling amphibians. (Modified from Thomson, 1966.)

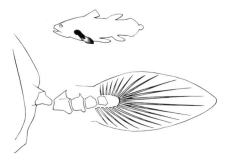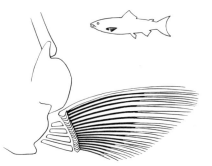

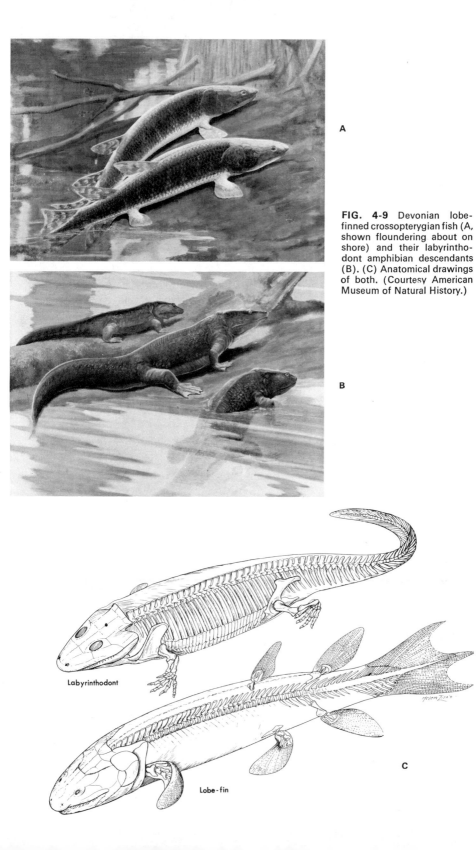

FIG. 4-9 Devonian lobe-finned crossopterygian fish (A, shown floundering about on shore) and their labyrinthodont amphibian descendants (B). (C) Anatomical drawings of both. (Courtesy American Museum of Natural History.)

A

B

Labyrinthodont

Lobe-fin

C

underside of the body. Members of one group, the "ray-finned" bony fishes, were by far the most successful as swimmers, and this group includes almost all present-day bony fishes (Fig. 4-8). Those in the other group, the "lobe-finned" fishes, were abundant only in Devonian time. From our point of view, however, they were far more important than their ray-finned cousins, for they gave rise to the land-dwelling amphibians and, through them, to all other groups of land-living vertebrates.

AMPHIBIANS—TRANSITIONAL VERTEBRATES

The Devonian Period was not only a critical time in the evolutionary history of fishes, but was also the time when vertebrates first made the transition to land, for the oldest fossil amphibians are found in Upper Devonian rocks of Greenland and eastern Canada. In adapting to life on land these early vertebrates faced the same problems of reproduction, water retention, and oxygen respiration that were earlier solved in plants and invertebrate animals. In addition, however, they faced a unique problem of locomotion on land. Plants, of course, do not move around, and the first land-dwelling invertebrates were arthropods and snails that developed from benthonic ancestors that were already adapted for moving on the solid surface of the sea floor by means of legs or a muscular foot. Fishes, on the other hand, were adapted to a swimming mode of life; a profound modification was required of their land-dwelling descendants.

It is the potential for developing a means of locomotion on land that distinguished the early lobe-finned fishes from their ray-finned relatives (Fig. 4-8). The bone arrangement of the former permitted the fins to move freely at their point of attachment to the body; in addition, the muscles extended *into* the fin to allow precise control of fin movements. Ray-finned fishes, on the other hand, lacked this flexible bone and muscle arrangement. The lobe-finned pattern, with its complex muscles, was ideally suited to develop into elongated, flexible limbs to support and move the animal on land.

The lobe-finned fishes also had another adaptation that fitted them for life on land. Most of them lived in freshwater streams and ponds that frequently became stagnant or even dried up completely during dry seasons. Lobe-fins (and some ray-fins and placoderms as well) were able to cope with this lack of water because they evolved auxiliary lungs, so that they could obtain oxygen directly from the air when the water became deficient in oxygen. We know that early lobe-finned fishes had lungs because they have some surviving air-breathing descendants—the "lungfishes" that live in streams and ponds in Africa, Australia, and South America. Like their early ancestors, these fishes breathe air when the water becomes stagnant. The presence of lungs in these living and fossil lobe-finned fishes was, like the fin arrangement, an adaptation that was to make possible vertebrate life on land.

The surviving lungfishes represent one group of lobe-finned fishes, but it was a second, less specialized group, the *crossopterygians*, that gave rise to the land-dwelling amphibians. Indeed, the skull and skeleton of Devonian crossopterygians are almost identical with those of the earliest fossil amphibians except for the modification of the lower fins into stubby limbs for walking on land (Fig. 4-9). The similarities are so close, in fact, that the crossopterygian-amphibian transition is one of the best documented in the entire fossil record. Although they flourished until the close of the Cretaceous Period, crossopterygians are unknown from Cenozoic rocks and were long thought to be extinct. In 1938, however, a South African fisherman hauled up a living specimen, which was later named *Latimeria* (Fig. 4-10). Since then several dozen individuals have been caught, and study of these amazing "living fossils" has helped to clarify the structures of the Paleozoic crossopterygians that gave rise to the amphibians.

FIG. 4-10   The "living fossil" crossopterygian fish *Latimeria*. The lobe-finned crossopterygians were thought to have become extinct in Cretaceous time until a living specimen was caught by a South African fisherman in 1938. Since then, several dozen have been obtained. The specimen shown here, about three feet long, was caught in 1960 and preserved by freezing. (Courtesy Peabody Museum of Natural History, Yale University.)

Probably the greatest remaining question is why these early fishes found it advantageous to develop limbs for moving about on land. It has been suggested that land locomotion first evolved as a means of searching for new streams and ponds during dry seasons, or as a way of escaping aquatic predators. Food might also have been more abundant on land. Probably all three factors were important, but, in any event, it is certain that the ability to move on land was a most useful adaptation, for amphibians rapidly expanded and diversified during the Carboniferous Period following their origin in Late Devonian time (Fig. 4-5).

Although amphibians solved the problems of air breathing and locomotion on land, they did not conquer the final difficulty of land life—reproduction. All amphibians (except for some very specialized present-day forms) must return to the water to reproduce, for their delicate eggs quickly dry out and die if deposited in the air. It remained for their descendants, the reptiles, to solve the reproduction problem by developing eggs with tough outer coverings to prevent drying.

FIG. 4-11 *Cacops*, an early Permian labyrinthodont amphibian. These amphibians had short, but very powerful, limbs and probably were completely terrestrial as adults, although the larvae were probably aquatic. Other, smaller tetrapods and insects furnished food. (Skeleton courtesy Field Museum of Natural History; Robert Bakker reconstruction.)

Surviving modern amphibians (frogs, toads, newts, salamanders, and their relatives) originated in Triassic and Jurassic time as specialized relatives of early amphibians, called *labyrinthodonts*, which dominated during the Carboniferous Period (Fig. 4-11). Labyrinthodonts were more reptilelike than their modern relatives, and, in general, looked something like fat, stubby-nosed alligators. The habits of many of them were probably alligatorlike as well, for they appear to have spent much of their lives in the water and along the banks of rivers and lakes. Throughout the Carboniferous Period the labyrinthodonts were the dominant land vertebrates, and some developed into large animals 10 feet long. They were still abundant in the Permian Period, but were beginning to decline as their more successful offspring, the reptiles, expanded. By the close of the Triassic Period, all the labyrinthodonts were extinct.

We shall return to the history of reptiles and their origin from the labyrinthodonts in Chapter 6, after considering, in the next chapter, the history of land plants, on which all land-dwelling vertebrates must ultimately depend for food.

five

land plants

In the last chapter, when we briefly traced the transition of aquatic plants to life on land, we saw that land plants of the Phylum Tracheophyta probably arose from green algae by developing root systems for gathering water and nutrients from the soil, leaves for photosynthesis in the air, and, most characteristic of all, a stem and a *vascular system* for transporting water and nutrients between the leaves and roots. This vascular system is the common characteristic of the Phylum Tracheophyta, which includes most land plants. As we saw, only the mosses and liverworts of the Phylum Bryophyta have adapted to land life without this system; they survive by remaining small plants confined to moist environments.

Fossil tracheophytes are first known from Late Silurian rocks and, since that time, the phylum has gone through three great periods of evolutionary expansion, each leading to progressively more successful groups of land plants. All three groups had well-developed roots, leaves, and vascular systems, but differed in the efficiency of their adaptations for reproduction on land. The first great expansion led to ferns and related plants that were to dominate the land through much of the Paleozoic Era. These groups had no seeds and lacked fully effective mechanisms for protecting the sperm from drying out; they were therefore still restricted to environments that were relatively moist. Beginning in the Carboniferous Period, these early seedless plants were gradually replaced by conifers and other groups that are known informally as *gymnosperms*. Gymnosperms differed from their predecessors by developing seeds and pollen, which are primarily adaptions to protect the sperm and insure successful reproduction in dry environments. These adaptations were highly advantageous, and by

Triassic time gymnosperms had replaced seedless plants as the dominant element of the land flora. Gymnosperms were the principal large land plants through much of the Mesozoic Era, and still today make up great forests of pine, spruce, and fir trees. They were to be overshadowed, however, by a third and final group of land plants, one that developed another adaptation for still more successful reproduction on land—the flower and its enclosed, covered seed. Fossil flowering plants first appeared in Lower Cretaceous rocks and then went through a rapid radiation so that by Late Cretaceous time flowering plants were dominant over much of the land surface. Today, aside from ferns and conifers, most familiar land plants belong to this group.

FOSSILIZATION OF LAND PLANTS

Before we look at the three plant groups that have successively dominated the land, it will be useful to review briefly the manner of preservation of fossil land plants. Unlike most of the animal phyla and many of the phyla of algae, vascular plants do not deposit calcium carbonate or other minerals that are readily preserved as fossils. In addition, many live in upland environments characterized by erosion rather than sedimentation; such forms are rarely buried in sedimentary deposits. For these reasons, the fossil record of land plants is less complete than is that of calcified algae and most of the animal phyla. Fortunately, however, the tough woody tissues of larger land plants are some-times preserved as *compressions*, which are flattened carbon-rich fossils formed when the weight of the overlying sediments drives off the volatile fluids of the plant, leaving only charcoallike remains of the original tissues (Fig. 5-1(B)). The most familiar such compressions are coal deposits (merely thick accumulations of ancient plant remains), but individual plant compressions are also common in many nonmarine sedimentary rocks. The great majority of fossilized vascular plants are preserved in this way, and much of our knowledge of fossil plants has come from the study of compressions.

Under ideal conditions still more revealing kinds of preservation can occur. Some of the earliest vascular plants are well known through a discovery, in 1917, of exceptional plant remains in a Lower Devonian chert near Rhynie, Scotland. When this "Rhynie Chert" was looked at in thin translucent slices under high magnification, it showed extraordinarily well-preserved stems and leaves of early land-dwelling tracheophytes, just as much older cherts have more recently yielded well-preserved Precambrian algae and bacteria. In these instances the fossils are not compressions, but undistorted and relatively unaltered remains that have been protected by becoming embedded in the hard, impermeable chert. Porous plant structures, especially woody stems, can also be preserved by becoming filled with silica, calcium carbonate, or other minerals to become the familiar "petrified wood" (Fig. 5-1(A)). Sometimes it is possible to reveal ex-tremely fine details of the original plant structure in such petrifactions by remov-

A

B

FIG. 5-1 Preservation of fossil plants. (A) Petrification; mineralized tree trunks of Eocene age. These trees were originally buried by volcanic debris, which has been eroded away to reveal the upright trunks. (Courtesy National Park Service.) (B) Compression; carbon-rich, compressed remains of the leaf of a Carboniferous seed fern. (Courtesy Theodore Delevoryas, 1966.) (C) Fossil spores and pollen; from Triassic rocks (magnified 375 times). (Courtesy Theodore Delevoryas.)

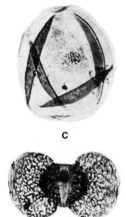

C

ing the minerals with acid, and by examining thin, polished slices under high magnification. Finally, acid treatment of shales and other fine-grained sedimentary rocks often yields fossil *spores* and *pollen*, specialized microscopic reproductive cells of vascular plants that are extremely tough and resistant to decay (Fig. 5-1(C)). From the combination of evidence from these modes of preservation, we now have a reasonably good knowledge of the history of the larger, woody land plants. The fossil record of smaller shrubs and herbs, which lack woody tissue, is less complete.

SEEDLESS FLORAS OF THE PALEOZOIC ERA

In the Classification of Organisms, p. 156, you will note that the Phylum Tracheophyta is divided into five subphyla; the first four are "seedless" vascular plants in contrast to the Subphylum Spermopsida, which includes all seed-bearing plants. Seedless vascular plants arose in Late Silurian time and dominated the land during the Devonian Period. They also made up a major element in the land floras throughout the rest of the Paleozoic Era, although some of the first primitive groups of seed plants were also common during the Carboniferous and Permian Periods (Fig. 5-2). Three of the four seedless groups—the Subphyla Psilopsida, Lycopsida, and Sphenopsida—declined rapidly after the Paleozoic Era, although all three groups still have a few surviving members. The Subphylum Pteropsida, which includes all the familiar modern ferns, also declined in importance after Paleozoic time, yet ferns remained relatively common after the expansion of more advanced plants.

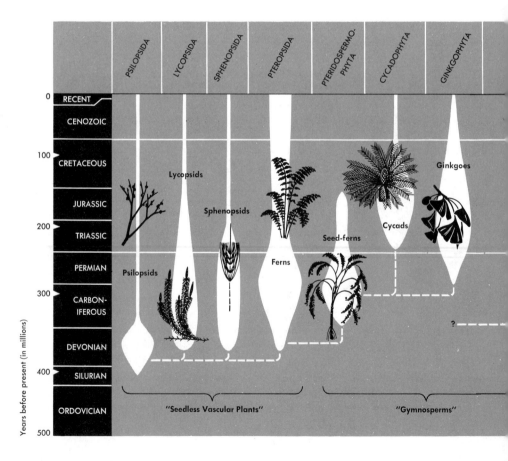

FIG. 5-2 Evolutionary history of the vascular plants. The dashed lines show the most probable evolutionary relations of the groups. The width of the white areas indicates the approximate abundance of each group.

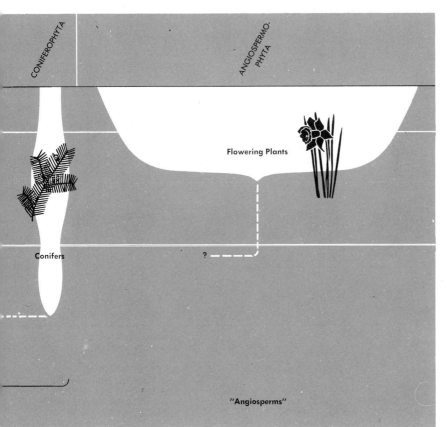

Reproduction in Seedless Plants

The four subphyla of seedless vascular plants differ in the nature and arrangement of the stems and leaves and in other aspects of their anatomy, but all share a primitive mode of reproduction (Fig. 5-3). The eggs and sperm are produced not by a large, leafy plant, but by a separate, small, leafless plant called a *gametophyte*. The sperm produced by the gametophyte is mobile; that is, it must *swim* a short distance to fertilize the eggs that are produced on another nearby gametophyte (in some groups each small gametophyte plant produces both eggs and sperm, but at different times and on different parts of the plant, to avoid self-fertilization; in other groups there are separate male and female gametophytes). Because these mobile sperm can move only through a film of water, seedless plants are generally restricted to moist environments. After fertilization, the egg produced by the small gametophyte develops into the familiar large leafy plant; this plant is called a *sporophyte* because at maturity it produces tiny *spores*, which are special reproductive cells that *do not require fertilization*. Spores develop on the leaves (the brown spots under the leaves of many ferns are spore-bearing structures) or on special stalks, and are scattered by the wind. When they fall on suitable moist ground they develop into the small gametophyte plant, which begins the reproductive cycle all over again. Among the vascular plants, this alternation of independent gametophyte and sporophyte plants is confined to the four seedless subphyla.

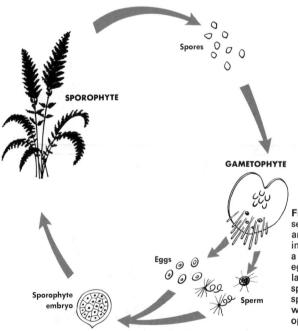

FIG. 5-3 Life history of a seedless vascular plant. Eggs and sperm are produced by an independent small plant called a gametophyte. The fertilized egg develops into the typical large plant, which is called a sporophyte because it produces spores, specialized cells from which the gametophyte develops to begin the cycle again.

FIG. 5-4 Seedless vascular plants of the Subphylum Psilopsida. (Left) A modern psilopsid found living today in the tropics (about one-fourth natural size). (Courtesy Arthur Cronquist.) (Right) Reconstruction of a Devonian psilopsid, based on fragments from the Rhynie Chert of Scotland. The swellings at the top are spore-producing structures. Note the absence of roots and leaves; photosynthesis took place in the upright stems and the horizontal stem at the base served to absorb water and nutrients.

Psilopsids

The most primitive of the four subphyla of seedless plants are the psilopsids. Psilopsids are the simplest vascular plants and, indeed, they have many algaelike characteristics. Like algae, most of them lack both roots and leaves. Photosynthesis takes place in the stem, and horizontal portions of the stem run along the ground to serve the function of roots. Psilopsids differ from algae, however, in having a simple vascular system. Psilopsids are the oldest fossil vascular plants for they first appear in Upper Silurian deposits; they are particularly abundant and diverse in rocks of the Devonian Period. The structure of these early psilopsids is well understood because they are exceptionally preserved in the Lower Devonian Rhynie Chert of Scotland (Fig. 5-4). The other three subphyla of seedless plants appear in the fossil record only shortly after the psilopsids, and the very simple structure of the latter suggests that they were ancestral to the others even though no transitional fossils are known. There are still a few living psilopsids, but the group has never been common since Devonian time.

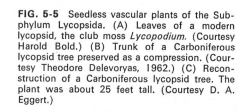

FIG. 5-5 Seedless vascular plants of the Subphylum Lycopsida. (A) Leaves of a modern lycopsid, the club moss *Lycopodium.* (Courtesy Harold Bold.) (B) Trunk of a Carboniferous lycopsid tree preserved as a compression. (Courtesy Theodore Delevoryas, 1962.) (C) Reconstruction of a Carboniferous lycopsid tree. The plant was about 25 feet tall. (Courtesy D. A. Eggert.)

A

C

B

Seedless Trees

The other three subphyla of seedless vascular plants differ from one another in general shape and form, but all developed leaves for efficient photosynthesis and root systems for gathering water and nutrients; all three groups include both small herbs and large, woody trees. Two of the subphyla, the Lycopsida (club mosses) (shown in Fig. 5-5), and the Sphenopsida (horsetails) (Fig. 5-6), contributed huge trees to the coal forests of the Carboniferous Period, but these seedless trees were overshadowed in Late Carboniferous and Permian time by the more efficient gymnosperms. The two subphyla survive today only as small, herbaceous plants. Ferns, too, had both herblike and treelike representatives in the Paleozoic forests, but both types have persisted (Fig. 5-7). The smaller ferns are familiar, but you may not realize that huge "tree ferns" are still found in many tropical forests (Fig. 5-7 left). These are the sole surviving trees of the great seedless forests of the Paleozoic Era (Fig. 5-8, p. 99).

A

B

FIG. 5-6 Seedless vascular plants of the Subphylum Sphenopsida. (A) A modern herbaceous Sphenopsid, the horsetail *Equisetum*. The segmented stem and leaf whorls are characteristics of sphenopsids. (B) Fossilized compression of a Carboniferous herbaceous sphenopsid. Sphenopsid trees were also abundant in Carboniferous forests, but only herbaceous forms survive today. (Both courtesy Field Museum of Natural History.)

FIG. 5-7 Seedless vascular plants of the Subphylum Pteropsida. (Left) A stand of modern tree ferns. Tree ferns are the only surviving seedless trees. (Courtesy Field Museum of Natural History.) (Right) Reconstruction of a Carboniferous tree fern. The plant was about 25 feet tall. (Courtesy J. Morgan.)

FIG. 5-8 Reconstruction of a Carboniferous forest dominated by seedless trees and seed ferns: lycopsids (L), sphenopsids (S), seed ferns (F). (Courtesy Illinois State Museum.)

EARLY SEED-BEARING PLANTS

With the decline of seedless Paleozoic plant groups, seed-bearing plants of the Subphylum Spermopsida came to dominate the land. The Spermopsida are divided into five classes; the first four (seed ferns, cycads, ginkgoes, and conifers) flourished in Late Paleozoic and Early Mesozoic time, but were largely replaced during the Cretaceous Period by the more advanced "flowering" plants of the final class, the Angiospermophyta (see Fig. 5-2, p. 93).

Reproduction in Seed-bearing Plants

Like the seedless plants before them, the four nonflowering classes of seed plants differ considerably in form and structure, but are united by similarities in their pattern of reproduction. In all four classes the seed is not enclosed by fleshy tissue, as it is in all flowering plants (for example, peas enclosed in a pod, apple seeds enclosed within the fruit) but is borne exposed, as on the woody scales of a pine cone. For this reason, the four classes have in the past been lumped into a single class, the "Gymnospermae" ("naked seed") in contrast to the Angiospermophyta ("covered seed"). Modern research has shown, however, that the four groups of gymnosperms are not closely related and are better separated as distinct classes comparable to the angiosperms. The term *gymnosperm* is still useful, however, as an informal way to refer to these four nonflowering, seed-bearing classes.

The gymnosperms and flowering plants had a great advantage over their seedless ancestors because they could reproduce without external moisture. Seed-bearing plants retain the special gametophyte stage found in seedless plants, but not as an independent plant living in moist soil. Instead, small male and female gametophyte plants develop within the moist tissues of the large sporophyte plant. The female develops into seed-producing organs and the male into a pollen grain, an amazingly specialized structure for preventing the sperm from drying during fertilization. Pollen grains, which are extremely tiny and tough, are carried by the wind or mobile animals to the reproductive organs of other plants. There they grow a moist tube that reaches to the eggbearing structure of the female gametophyte. The sperm then pass through this tube to fertilize the egg, thus insuring a continuously moist environment. Pollen grains not only overcame the problem of protecting the sperm from drying, but also provided a means of readily transporting the sperm to distant plants for cross-fertilization. After fertilization, the egg develops into an embryonic sporophyte plant surrounded by starchy nutrients; the whole being a seed with one or more protecting coats investing the embryo and its food store. When transported to a favorable environment by wind or animals, the seed sprouts to become a new sporophyte plant (Fig. 5-9).

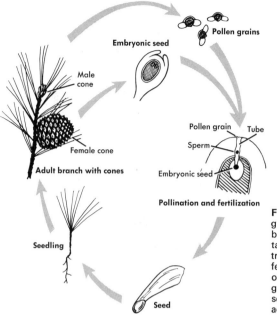

FIG. 5-9 Life history of a coniferous gymnosperm. Male cones produce sperm-bearing pollen grains; female cones contain embryonic seeds. When pollen is transported by the wind to a female cone, fertilization is accomplished by movement of the sperm through a moist tube which grows from the pollen grain. The fertilized seed then grows into the cone-bearing adult.

Seed Ferns

The earliest gymnosperms are the seed ferns of the Class Pteridospermophyta, an extinct group with characteristics of both the seedless ferns and the seed-bearing Subphylum Spermopsida. The leaves of these plants are so fernlike that most of them were for many years identified as ferns (Fig. 5-10). Gradually, however, seed-bearing structures were discovered in association with the leaves, and it is now apparent that many early fossil "ferns" are really seed-bearing gymnosperms. The oldest known seed ferns are found in Lower Carboniferous rocks, and they are common, along with true ferns, lycopsids, and sphenopsids, throughout the Carboniferous and Permian Periods. They declined rapidly, however, during Early Mesozoic time as the more advanced cycads, ginkgoes, and conifers expanded to dominance. Fossil seed ferns are last known from Jurassic rocks; they are the only major division of the vascular plants that is extinct. In many ways, the seed ferns played the same role for the seed plants that the earlier psilopsids played for the entire phylum of vascular plants. Both were early experimental groups that later gave rise to more advanced and successful plants.

FIG. 5-10 Gymnosperms of the Class Pteridospermophyta. (Right) Fossilized compression of a frond of a Carboniferous seed fern. (Courtesy Henry N. Andrews.) (Above) Reconstruction of a Carboniferous seed fern. The plant was about 15 feet tall. The seeds were borne in clusters on short, central stems. (Courtesy Theodore Delevoryas.)

Cycads and Ginkgoes

The next two classes of gymnosperms to develop, the Cycadophyta (cycads and their relatives) and Ginkgophyta (ginkgoes), were the dominant trees of the Triassic and Jurassic Periods, but they declined rapidly after the Cretaceous expansion of flowering plants and are relatively uncommon today. The cycads include two distantly related groups with similar form and leaves, but with different seeds and reproductive structures. Both are superficially palmlike trees (true palms are flowering plants); one group became extinct in Cretaceous time, but a few genera of the other group survive today in warm regions (Fig. 5-11).

A

FIG. 5-11 Gymnosperms of the Class Cycadophyta. (A) Fossilized leaf compression of a Mesozoic cycad. (Courtesy Theodore Delevoryas.) (B) A modern cycad forest. (Courtesy Field Museum of Natural History).

B

The ginkgoes, which have a characteristic fan-shaped leaf (Fig. 5-12), have survived as a single species that was originally native to Asia. Wild individuals are now thought to be extinct, but the tree thrives in cultivation and is commonly planted as a shade tree in the eastern United States and in other temperate regions. Ginkgo trees thus provide a rare example of an organism that has been preserved, rather than eliminated, by the actions of man.

FIG. 5-12 Gymnosperms of the Class Ginkgophyta. Fossilized leaf compression of a Miocene ginkgo, with modern leaves below for comparison. The fan-shaped leaves are characteristic of the group.

Conifers

Both cycads and ginkgoes probably arose from a seed fern ancestor, but the origin of the remaining gymnosperm group, the familiar conifers with their cones and needlelike leaves, is uncertain. They first appear in the Carboniferous Period along with less advanced seedless plants and seedferns. For this reason, and because of certain structural peculiarities, many botanists feel that they may have evolved separately from seedless ancestors, rather than originating from a seed fern as did the cycads and ginkgoes. The Paleozoic conifers are mostly rather primitive, extinct forms; the familiar modern groups, such as the pines, firs, and cedars, arose during Early Mesozoic time, when conifers expanded along with the cycads and ginkgoes. They survive today as the dominant forest trees in areas where water is scarce, soils are poor, or winters are cold. There they provide our closest modern analogue to the great conifer-cycad-ginkgo forests that covered the land during the Triassic and Jurassic Periods.

THE RISE AND DOMINANCE OF FLOWERING PLANTS

The final group of vascular plants, the flowering plants of the Class Angiospermophyta, are among the most diverse and successful organisms that have ever evolved. Today they overwhelmingly dominate the land: of the 260,000 or so living species of vascular plants, about 250,000, or 96 percent, are angiosperms (the other 10,000 species are mostly ferns; gymnosperms, despite wide distribution and local importance, have only 700 surviving species).

The great evolutionary success of flowering plants is partly due to the development of the flower and its enclosed seed (Fig. 5-13). Flowers are specialized reproductive structures that normally bear both the seed-producing and pollen-producing organs, surrounded by modified, colored leaves (the petals and related structures), which serve principally to attract animal pollinators. The flowerless gymnosperms rely on the wind to carry the pollen from plant to plant, a relatively wasteful and inefficient means of fertilization. In contrast, the angiosperm flower is basically an adaptation that insures fertilization by attracting insects or birds to transport the pollen.

Like the flower, the enclosed angiosperm seed is a reproductive advance over the naked seed of the gymnosperms. In particular, the enclosed seed permits the development of fleshy, edible fruit coverings that insure seed dispersal when they are eaten by animals. Angiosperms have also developed a host of other seed dispersal mechanisms, such as the tufted, wind-transported fruit of the dandelion and prickly burs that adhere to the surface of passing animals. These adaptations for achieving wide distribution of the seeds are all made possible by the enclosed structure of the angiosperm seed.

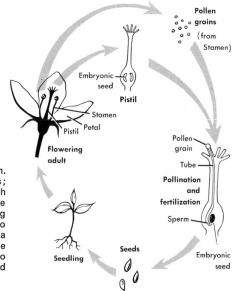

FIG. 5-13 Life history of an angiosperm. Pollen is produced in the flower by stamens; seeds develop in the pistil (in some plants both are found in the same flower; in others there are separate stamen-bearing and pistil-bearing flowers). When a pollen grain is transported to the pistil, fertilization takes place through a moist pollen tube as in the gymnosperms. The colored petals of the flower serve principally to attract animal pollinators, especially birds and insects.

Flowering Plant Diversity

Although much of the evolutionary success of the angiosperms is undoubtedly due to their mode of reproduction, probably equally important were some poorly understood factors that led to extraordinary structural plasticity and permitted the development of a wide variety of shapes and habits. Three of the four subphyla of seedless plants evolved both small herbs and shrubs as well as large trees. Present-day ferns show all of these forms, but only small herbaceous lycopsids and sphenopsids have survived, the larger trees having become extinct. In contrast, the gymnosperms always seem to have been large trees and woody shrubs; there is no evidence that the group has ever developed small herbs. Ferns and the other seedless groups, then, must have been the principal small, nonwoody plants not only during the Paleozoic Era, but also into the Triassic and Jurassic Periods when gymnosperms were the dominant forest trees. In striking contrast to the gymnosperms, the angiosperms have diversified into almost every imaginable plant habit.

FIG. 5-14 Flowering plants of the Class Angiospermophyta. Three fossilized leaf compressions of Eocene and Miocene flowering trees. (Courtesy Henry N. Andrews.)

The earliest angiosperms were probably woody trees or shrubs (Fig. 5-14), but the group rapidly developed a host of smaller herbs. Some of these, such as the cacti and their relatives, have adaptations for water storage and survival in arid regions. Others have gone to the opposite extreme and become adapted for life in streams, ponds, and lakes. A few grasslike forms have even reinvaded the seas to become common in shallow coastal waters. Still others, such as common mistletoe, have become adapted for a commensal life on larger plants. Some are even outright parasites that use the tissues and fluids of the host plant as nutrients. A bizarre adaptation occurs in insectivorous plants, whose flowers or leaves trap insects for nourishment. These diverse adaptations illustrate the structural plasticity and evolutionary success of flowering plants.

The Origin of Flowering Plants

One of the most perplexing problems in the entire evolutionary record of plants concerns the origin of the angiosperms. Scattered fossil fragments that may represent angiosperms are occasionally found in Triassic and Jurassic rocks, but the certain record of the group begins suddenly in Lower Cretaceous deposits. Moreover, many of the major angiosperm subgroups are already differentiated when they first appear in the fossil record. This fact suggests that the group had a considerable pre-Cretaceous history, which, for some reason, is not recorded in the fossil record. Perhaps these earlier angiosperms are unknown because they were rare and local in occurrence, or were confined to highland areas where erosion was active and fossilization unlikely. In any case, there are no transitional fossils to indicate the ancestry of the group. Possibly they originated from the seed ferns or from some other group of gymnosperms, but some botanists think they may have evolved directly from seedless ancestors.

Whatever their origin, angiosperms went through an extremely rapid evolutionary radiation during Cretaceous time, and by the end of the period they had replaced the gymnosperms as the dominant plants over much of the land surface. This radiation continued throughout the Cenozoic Era to produce the staggering diversity of flowering trees, shrubs, vines, and herbs that cover the land today.

six

reptiles and mammals

Now that the evolutionary record of land plants has been reviewed, we can return to the land-dwelling vertebrates, all of which ultimately depend on land plants for their nourishment. In Chapter 4 we traced the history of the earliest land vertebrates—primitive amphibians that arose from air-breathing, lobe-finned fish. There we saw that amphibians have never fully adapted for land life because they lack a means of protecting the embryo from drying out and thus must return to water to lay their eggs. This problem was overcome in the reptiles, a group that arose from the early amphibians. Reptiles, in turn, were the ancestors of the two most successful classes of land vertebrates, the birds and mammals.

Because birds are such a large and important group of land vertebrates, why is this chapter not called "Reptiles, *Birds*, and Mammals"? The reason is that birds have by far the poorest fossil record of any of the vertebrate classes. A few rare fossils provide important clues about the history of the group, but in contrast to the relatively complete evolutionary record of reptiles and mammals, we have almost no direct knowledge of bird evolution.

The scarcity of fossil birds re-emphasizes the two essential requirements for fossilization that were discussed in Chapter 3: (1) an animal or plant must have had hard parts capable of preservation, and (2) it must have lived in an environment where it was likely to become rapidly buried after death. All land vertebrates have bones that fulfill the first requirement, but, because of the second, they are not equally common as fossils. In general, marsh- and swamp-dwelling land vertebrates are the most frequently fossilized, often being found as whole, articulated skeletons. Upland and tree-dwelling vertebrates are much rarer as whole skeletons, and even complete skulls are relatively uncommon. Instead, such forms are usually preserved only as isolated jaws and teeth. Teeth, in particular, are common because they are extremely hard and resist abrasion and decay much longer than bone. Flying vertebrates are among the rarest of fossils because their skeletons are extremely fragile and are almost always exposed to decay and predation when they fall to the ground after death. The few rare fossils of flying forms have generally been found in lake or marine deposits, indicating that the animals somehow died in or above the water, sank to the bottom, and were rapidly buried by sediment.

From this brief discussion you can appreciate that fossil land vertebrates, particularly whole skeletons and skulls, are relatively much less common than are fossils of shelled invertebrates and mineralized algae, which are ordinarily preserved as whole individuals and in tremendous numbers. For this reason, vertebrate paleontologists frequently must search at favorable sites for many years to obtain even a few good specimens. These efforts have, however, produced impressive results, for we now know more about the evolutionary history of reptiles and mammals than about any other groups of organisms.

EARLY REPTILES

Primitive amphibians can be readily distinguished from their fish ancestors because they acquired new skeletal structures, the limbs, that are recognizable in fossil skeletons. In contrast, the transition between amphibians and their reptile descendants is less clear from the fossil record because the principal differences between amphibians and reptiles are found not in skeletal features but in mode of reproduction. Reptiles overcame the problem of reproduction on land with the shelled egg, a device for allowing the embryo to grow in its own self-contained liquid environment. After development is completed, the animal breaks out of the shell as a small but otherwise fully formed animal. Many early reptiles retained the semiaquatic habits of their amphibian ancestors, and thus it is possible that the shelled egg, which must be deposited on land, first evolved as a means of avoiding loss of the larval stages to fish and other aquatic predators. Nevertheless, the reptilian egg also freed land vertebrates from the necessity of living near large bodies of water and ultimately permitted them to wander freely over the land surface in search of food and favorable habitats.

Although it is impossible to determine directly from the skeletal features whether or not a given fossil vertebrate layed shelled eggs, there are some minor skeletal differences, principally in the patterns of skull bone and vertebral construction, that distinguish amphibians and reptiles, and these features are useful in separating fossil representatives of the two groups. Such evidence shows that by late in the Carboniferous Period the first reptiles had evolved from closely similar amphibian ancestors. During the succeeding Permian Period reptiles went through an explosive evolutionary radiation that began their long domi-

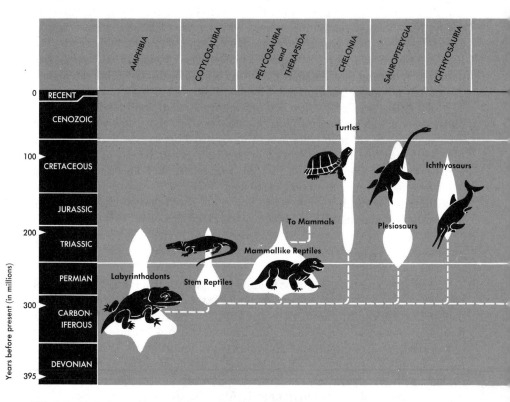

FIG. 6-1 Evolutionary history of the reptiles. The dashed lines show the most probable evolutionary relations of the groups. The width of the white areas indicates the approximate abundance of each group.

nance of the land, a dominance that was to last until the end of the Mesozoic Era (see Fig. 6-1). As the numbers of reptiles expanded in the Permian Period, the less efficient amphibians began to decline in importance. Labyrinthodont amphibians were extinct by the close of the Triassic Period; since Triassic time the principal amphibians have been frogs and salamanders, which have survived to the present day by remaining small and by leading unobtrusive lives around lakes, rivers, and streams.

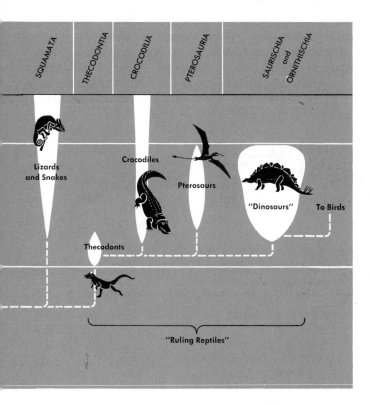

Cotylosaurs and Mammallike Reptiles

Labyrinthodonts were quite unlike modern frogs and salamanders but, instead, superficially resembled fat, stub-nosed alligators. The first reptiles, similar-looking forms called *cotylosaurs* (Fig. 6-2), evolved from the labyrinthodonts. The cotylosaurs, in turn, gave rise to all other reptile groups, and for this reason they are commonly referred to as *stem reptiles*. The reptilian descendants of the early cotylosaurs were to rule the land in two successive phases. During Permian and much of Triassic time the dominant land vertebrates were the *mammallike reptiles*, an abundant and diverse group that was to give rise to the mammals. In Late Triassic time the mammallike reptiles were replaced by a more familiar group of land reptiles—the dinosaurs—that were to be masters of the land throughout the Jurassic and Cretaceous Periods. The importance of the dinosaurs in reptilian history has long been appreciated, but only recently have paleontologists begun to emphasize the great abundance and diversity of the mammallike reptiles that preceded the dinosaurs and dominated the land for almost as long.

FIG. 6-2 A Permian cotylosaur, or "stem reptile." (Above) Skeleton of an Early Permian cotylosaur, *Labidosaurus,* about two feet long. Cotylosaurs are the ancestral group from which all other reptiles arose. (Courtesy Field Museum of Natural History.) (Below) Restoration of *Limnoscelis,* a Late Carboniferous relative of *Labidosaurus.* One of the most primitive cotylosaurs known, *Limnoscelis* was about five feet long and had slow-acting, large and powerful limbs for fully terrestrial locomotion. (Robert Bakker reconstruction.)

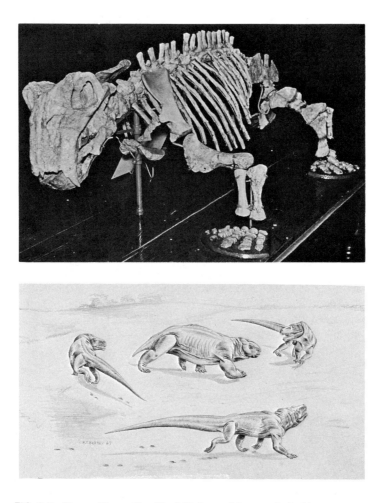

FIG. 6-3 Mammallike reptiles. (Top) Skeleton of the aquatic, herbivorous, mammallike reptile, *Lystrosaurus,* about three feet long, from Early Triassic rocks. The sprawling posture characterized all mammallike reptiles and even early mammals. (Courtesy A. W. Crompton.) (Bottom) An Early Triassic scene in China. In the center is *Sinokannemeyeria,* one of the larger relatives of *Lystrosaurus.* These mammallike reptiles had toothless beaks for slicing plants. Attacking from three sides is *Shansisuchus,* an agile, powerful, carnivorous thecodont, about five feet long. (Robert Bakker reconstruction.)

In spite of their name, most mammallike reptiles did not look very much like mammals. Most were large-headed, bulky animals from 2 to 15 feet long that must have moved rather awkwardly, for the legs were not directly beneath the body, as in most dinosaurs and mammals, but instead extended outward from the side of the body (Fig. 6-3). Skulls and teeth show that both herbivores and predatory carnivores were common. During the Triassic Period several different lines developed progressively more mammalian skull features, and, as we shall see, one of these gave rise to the mammals in Late Triassic time.

Turtles, Lizards, and Snakes

Throughout most of the Permian Period, large mammallike reptiles, along with their surviving forebears, the cotylosaurs and labyrinthodonts, were the principal land vertebrates. Near the beginning of the Triassic Period several more specialized reptilian groups arose from the stem cotylosaurs. The turtles and lizards (including their close relatives, the snakes), are familiar because they are the principal present-day reptiles. Turtles have changed very little since they first arose in Early Triassic time. Apparently their unique protective shell and aquatic, omnivorous habits allowed them to survive the crises that later eliminated most of their reptilian relatives. Lizards and snakes, although they are the most widespread and successful modern reptiles, are rather uncommon fossils. The first fossil lizards are found in Upper Triassic rocks; snakes are a late evolutionary offshoot from the lizards, first known from Cretaceous deposits. Snakes stand almost alone among the vertebrates in their development of toxic poison glands in many species, an adaptation that has contributed to their evolutionary success by making them an object of dread among other land vertebrates. Both snakes and lizards, although dominantly carnivores, are rather secretive and inconspicuous animals that hide from their enemies in dense foliage and rocks; this habit has undoubtedly been a strong factor in their long survival.

A

Marine Reptiles

Two other groups that arose from the cotylosaurs, the extinct ichthyosaurs and plesiosaurs, secondarily returned to the sea to become large, impressive swimming reptiles (Fig. 6-4). The sharklike ichthyosaurs reached a length of 12 feet, but some long-necked Cretaceous plesiosaurs were gigantic beasts 50 feet long. Both were predatory carnivores that fed mostly on fish; both groups originated in Triassic time and are last known from Cretaceous rocks. Today the only sea-dwelling reptiles are a few turtles, lizards, and sea snakes.

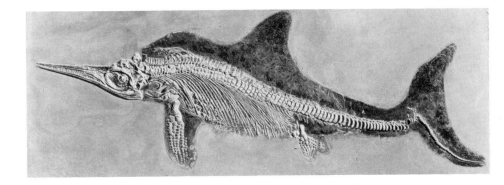

FIG. 6-4 Plesiosaurs and ichthyosaurs, large Mesozoic marine reptiles. (A, opposite page) Plesiosaur, mounted Jurassic fossil skeleton about 20 feet long. (B) Ichthyosaur, Jurassic fossil skeleton with the outline of the body preserved. The skeleton is about nine feet long. (Both courtesy American Museum of Natural History). (C) Drawing showing probable life appearance of plesiosaurs (left) and ichthyosaurs (right). (Knight reconstruction, courtesy American Museum of Natural History.)

C

DINOSAURS AND OTHER RULING REPTILES

Thecodonts, the remaining group that arose from the cotylosaurs, are of great evolutionary importance because they provided the root stock for a second great reptile radiation, that of the "ruling reptiles," which became abundant in Late Triassic time and were to dominate the land during the Jurassic and Cretaceous Periods.

Thecodonts

Thecodonts first appear in earliest Triassic time and almost immediately went through an evolutionary radiation that led to many adaptive types (Fig. 6-5). Some developed into crocodilelike aquatic animals, others were quadripedal land carnivores, and a few developed the unique habit of walking upright on their hind legs. This bipedal posture allowed them to move around more rapidly in search of their prey, for the bipedal forms were predatory carnivores. Mostly small animals, the thecodonts were generally overshadowed by their larger Triassic neighbors, the mammallike reptiles. Nevertheless, they had one adaptive advantage that was to have far-reaching evolutionary consequences. Recall that most mammallike reptiles had an awkward sprawling posture because the limbs extended outward from the sides of the body. Many thecodonts, on the other hand, had limbs that were under the body, allowing more efficient support of the body weight and, consequently, more effective mobility. Furthermore, in the bipedal forms the front limbs were no longer required for walking but were free to develop into new structures, such as grasping claws for holding prey or wings

FIG. 6-5 Restoration of *Euparkeria*, an Early Triassic thecodont reptile that could move bipedally when running fast. Thecodonts similar to *Euparkeria* were the ancestors of many important groups, including pterosaurs, crocodiles, and dinosaurs. (Robert Bakker reconstruction.)

to permit the conquest of the air. This more efficient posture is largely responsible for the success of the three groups of reptiles that arose from the thecodonts: the pterosaurs, which were flying reptiles, now extinct; the familiar surviving crocodiles and their relatives; and the dinosaurs, which include the largest and most awesome land animals ever to evolve.

Pterosaurs and Crocodiles

The ability to fly is an advantageous adaptation, for it permits an animal to escape easily from predators and to cover effortlessly a great deal of territory in search of food. Insects and birds have been outstandingly successful in exploiting this habit, yet the ability to fly has also evolved in reptiles and mammals. Shortly before birds developed the feathered wing, a group of true reptiles, the pterosaurs, were experimenting with flight by other means. They evolved one extremely long finger to which was attached a thin membrane of skin that functioned as the wing (Fig. 6-6). Like the dinosaurs, pterosaurs developed into giants; the biggest, with a wingspread of more than 50 feet, was the largest animal ever to fly. The bodies of these huge pterosaurs were, however, very small (about the size of a large bird), and it is unlikely that they had sufficient muscle power to move the wings actively during flight. Instead, they probably glided on rising currents of air, a habit that is characteristic of large-winged sea birds today. Like birds and other flying animals, pterosaurs are rare fossils. Many kinds developed during the Jurassic and Cretaceous Periods, but the group apparently could not compete successfully with birds, for they are last found in Upper Cretaceous rocks.

The crocodiles, another group that originated from the thecodonts, have apparently always been specialized, semiaquatic carnivores. As the only surviving reptilian descendants of the thecodonts, crocodiles are perhaps of greatest interest as close living relatives of the dinosaurs.

Dinosaurs

The most dramatic descendants of the thecodonts were the many dinosaurs that dominated Jurassic and Cretaceous landscapes. *Dinosaurs* is a popular term for two unrelated orders (Saurischia and Ornithischia) of large land reptiles; the two orders differed in features of skeletal structure that are not important here, and we shall refer to both informally by the familiar term *dinosaurs*. Dinosaurs made up the second of the three major groups of dominant land-dwelling vertebrates: mammallike reptiles during the Permian and Triassic Periods, dinosaurs during the Jurassic and Cretaceous Periods, and mammals since the close of the Cretaceous Period (Fig. 6-7). All three groups have shown similar tendencies to evolve many specialized herbivores and relatively few carnivores, which prey upon the herbivores. This trend is particularly well illustrated by the dinosaurs.

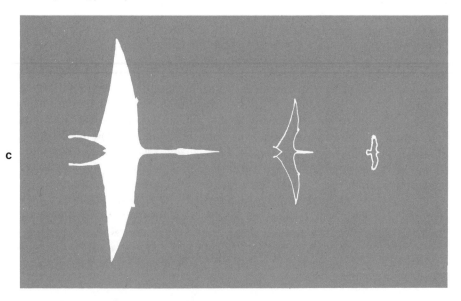

FIG. 6-6 Pterosaurs, large Mesozoic flying reptiles. (A) A complete Cretaceous fossil skeleton and restored body outline. This specimen had a wingspread of about 20 feet. A mounted condor, the largest living bird, is shown for comparison. (B) Drawing showing probable life appearance of the skeleton. (Both (A) and (B) courtesy of American Museum of Natural History.) (C) Comparative wingspan of pterosaurs. The specimen at left, the largest yet discovered, was recently found in latest Cretaceous deposits of West Texas. It has a probable wingspan of over fifty feet. The smaller pterosaur and the condor shown in (A) are outlined for comparison. (Modified from Lawson, 1975, *Science*, v. 187, p. 947.)

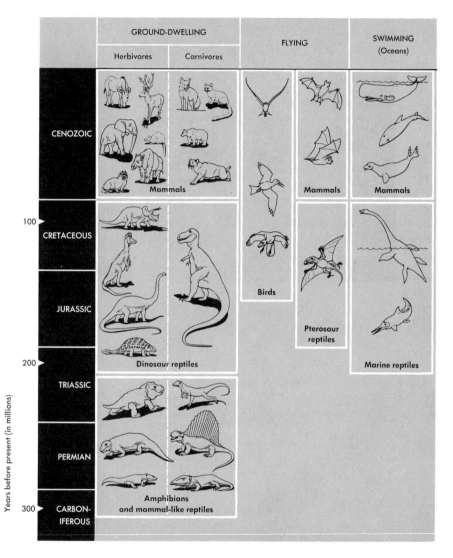

FIG. 6-7 Periods of dominance of the principal groups of vertebrates (other than fishes). There have been three dominant faunas of ground-dwelling vertebrates: (1) amphibians and mammallike reptiles in the Carboniferous, Permian, and Early Triassic Periods; (2) dinosaurs in the Late Triassic, Jurassic, and Cretaceous Periods, and (3) mammals during the Cenozoic Era. Each group has developed both herbivorous and carnivorous lines. The flying reptiles (pterosaurs) and marine reptiles of the Jurassic and Cretaceous Periods have been replaced by bats and marine mammals during the Cenozoic Era.

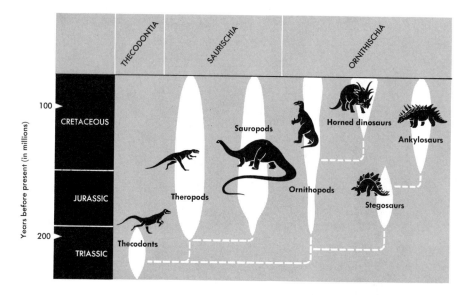

Years before present (in millions)

FIG. 6-8 (Above) Evolutionary history of the dinosaurs. The dashed lines show the most probable evolutionary relations of the groups. The width of the white areas indicates the approximate abundance of each group.

FIG. 6-9 (Below) Mounted skeleton of the Cretaceous theropod dinosaur, *Tyrannosaurus*, the largest known land-dwelling predator. The skeleton is about 18 feet high and 50 feet long. The probable life appearance of *Tyrannosaurus* is shown in Fig. 6-14. (Courtesy American Museum of Natural History.)

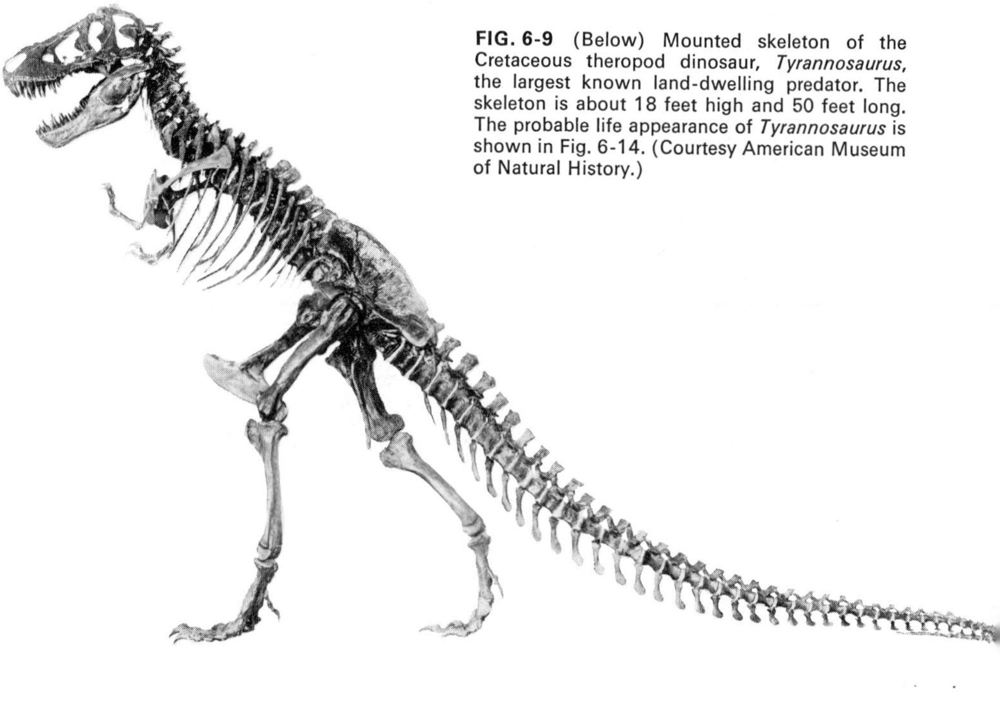

There were six principal kinds of dinosaurs. Studies of fossil skulls and teeth show that five of them were herbivorous—the *ornithopods, stegosaurs, ankylosaurs, horned dinosaurs*, and *sauropods*. The sixth group, the *theropods*, were carnivorous. Each of these six lines shows an evolutionary development from relatively small, unspecialized forms found in Upper Triassic and Lower Jurassic rocks to the familiar huge and impressive animals of Late Jurassic and Cretaceous time. Although the early dinosaurs were no less important, special attention will be directed to these large end products of each line, for they show most clearly the characteristic specializations of each group (Fig. 6-8). In the five herbivorous lines, evolutionary trends appear to have been directly related to defense against the theropods, which, in the familiar *Tyrannosaurus*, developed the most awesome flesh-eater to evolve (Fig. 6-9). Each of the five herbivorous groups developed means of protection from the theropods; some by becoming rapid runners, others by developing protective armor and weapons for defense, and still others by becoming too large for attack and by living in the waters of rivers and lakes.

All of the carnivorous theropods and one group of herbivores, the orni-thopods, were mobile, bipedal creatures (Fig. 6-10). Apparently, the long hind legs and upright posture permitted the ornithopods to escape from the theropods by running away, for they appear to have had no other means of protection.

Bony armor developed as a means of protection in two related groups: the bizarre stegosaurs, with their tiny heads and large bony plates running down the spine, and the ankylosaurs, low, short-legged forms with a more complete covering of bony armor (Figs. 6-11 and 6-12). In addition to the protective armor, both of these groups developed peculiar weapons at the end of the tail— long spikes in the stegosaurs and a heavy, bony, clublike structure in the anky-losaurs.

The greatest development of defensive weapons occurred in the fourth herbivorous group, the horned dinosaurs (Fig. 6-13). They had huge skulls with long, stout horns that must have been very efficient weapons for defense against the theropods.

The fifth group of herbivorous dinosaurs, the huge, long-necked sauropods, were the largest land animals ever to evolve (Fig. 6-14). They probably spent much time feeding on aquatic plants in rivers and lakes, where their huge bodies were buoyed up by water. Sauropods have no defensive skeletal features, but apparently their aquatic habit and huge size were sufficient protection against the theropods.

The six principal dinosaur groups did not originate at the same time. The sauropods, theropods, ornithopods, and stegosaurs evolved in Late Triassic or Early Jurassic time, whereas the ankylosaurs and horned dinosaurs developed in the Late Jurassic and Cretaceous Periods. It is a striking fact, however, that almost all of the dinosaur lines persisted until late in the Cretaceous Period and then rather abruptly became extinct. The only exception was the stegosaurs,

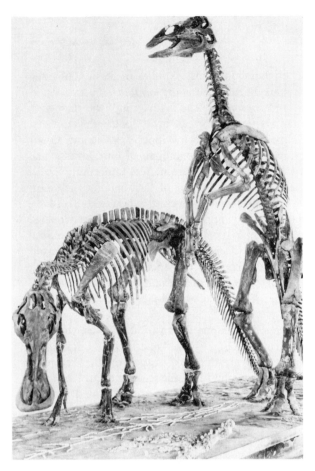

FIG. 6-10 Ornithopod dinosaurs. (At left) Mounted skeletons of the Cretaceous ornithopod *Anatosaurus*. Ornithopods were plant eaters that moved around rather swiftly on their hind legs. The skeletons are about 15 feet high. (Courtesy American Museum of Natural History.) (Below) Painting showing probable life appearance of the skeletons. (Knight reconstruction, courtesy American Museum of Natural History.)

which died out earlier, in Early Cretaceous time. Reasons for the Late Cretaceous demise of the dinosaurs, like the reasons for other widespread extinctions, are obscure and uncertain. Apparently, some environmental factor or factors changed so rapidly that dinosaur evolution could not keep pace. Whatever the cause, the extinction of the dinosaurs left vacant much of the land surface and thus paved the way for the great evolutionary expansion of the mammals, which were to be the dominant land vertebrates throughout the Cenozoic Era.

FIG. 6-11 Stegosaurs, armored plant-eating dinosaurs. (Right) Mounted skeleton of *Stegosaurus* from Jurassic rocks. The skeleton is about 17 feet long and 9 feet high. Note the peculiar armored plates along the backbone, the small head, and the spiked tail for defense against the predatory theropods. (Courtesy Smithsonian Institution.) (Below) Painting showing probable life appearance of the skeleton. (Knight reconstruction, courtesy American Museum of Natural History.)

FIG. 6-12 (Above) Reconstructed model of *Ankylosaurus*, an armored plant-eating dinosaur from Cretaceous rocks. The animal was about 15 feet long. Note the heavy, clublike tail for defense against predatory theropods. (Courtesy American Museum of Natural History.)

FIG. 6-13 Horned dinosaurs. (Above) Mounted skeleton of *Triceratops*, a horned plant-eating dinosaur from Cretaceous rocks. The skeleton is about 24 feet long. Note the huge, massive skull and long horns for defense against the carnivorous theropods. (Courtesy American Museum of Natural History.) (Below) Painting showing probable life appearance of the skeleton. The horned dinosaur is being attacked by the theropod *Tyrannosaurus*. (Knight reconstruction, courtesy Field Museum of Natural History.)

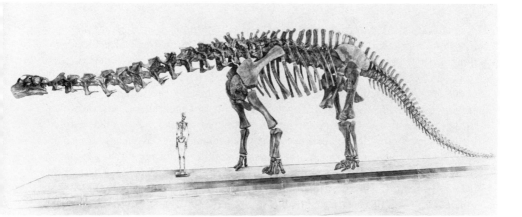

FIG. 6-14 Sauropod dinosaurs. (Above) Mounted skeleton of *Apatosaurus* (formerly known as *"Brontosaurus"*), a plant-eating sauropod from Jurassic rocks. The skeleton is about 70 feet long; a human skeleton is shown for scale. Sauropods are the largest known land animals. (Courtesy American Museum of Natural History.) (At right) Painting showing probable life appearance of the skeleton. (Courtesy Sinclair Refining Company.)

BIRDS

Biologists have long recognized close anatomical similarities between birds and reptiles. Unlike modern reptiles, however, birds are "warm-blooded;" that is, they have the ability to maintain the body at a constant temperature, an adaptation that marks an advance over many of their reptile ancestors, whose body temperature was determined by the temperature of the surrounding air. In their egg-laying reproduction and general anatomy, however, birds can be accurately characterized as "feathered reptiles." As we have noted, birds are rarely fossilized, yet among the most remarkable fossils ever found are several skeletons from Jurassic limestones of southern Germany that are clearly transitional between birds and their reptilian ancestors (Fig. 6-15). These skeletons have teeth (all modern birds are toothless) and other features that are so much like those of certain small dinosaurs that, had the skeletons alone been preserved, they probably would have been described as reptiles. Fortunately, the fine-grained limestone in which the skeletons are preserved also show clear impressions of flight feathers on the tail and on the elongated front limbs. Feathers, found only in birds, indicate that this strange animal was indeed a primitive bird that still retained the skeletal features of its reptilian ancestors. Regrettably, relatively few comparable bird skeletons are known from younger rocks, and these give a mere glimpse of the evolutionary history of the class.

MAMMAL ORIGINS

We have already seen that it is difficult to distinguish the earliest fossil reptiles from their amphibian ancestors because the two groups differed primarily in mode of reproduction, a feature not readily apparent in fossil skeletons. A similar situation exists between reptiles and their mammalian descendants, for the most significant differences are again reproductive and physiological rather than skeletal. Most importantly, mammals maintain a constant body temperature, whereas the body temperature of present-day reptiles is determined by the surrounding air temperature. This adaptation permits mammals to lead an active and diversified life, for they can survive in cold regions and can search for food in all seasons and during the cool of the night as well as in the warmth of the day.

A number of anatomical and physiological features are correlated with increased mammalian activity and temperature regulation. Insulating hair developed on the body surface as a means of conserving body heat, and a more efficient heart and lungs evolved for better oxygenation of the blood. Reptiles normally swallow their food whole and then remain inactive for long periods while digestion takes place. Active mammals, on the other hand, require smaller and more easily assimilated bits of food. This difference is reflected in the patterns of the jaws and teeth. All mammals, and some of their mammallike reptilian

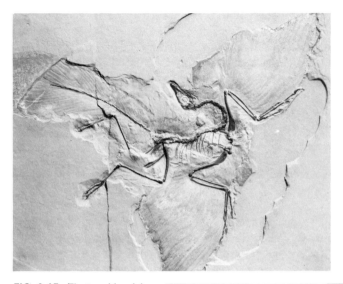

FIG. 6-15 The transitional Jurassic bird, *Archaeopteryx*. (Above) A nearly complete fossil showing the reptilelike skeleton as well as impressions of wing and tail feathers. The skeleton is about 12 inches long. (Courtesy American Museum of Natural History. (At right) Drawing showing the probable life appearance of the skeleton. (Rudolf Freund restoration, courtesy Carnegie Museum.)

forebears, developed several different kinds of teeth, such as molars and incisors, for seizing food and for chewing and grinding it before swallowing. In most reptiles, on the other hand, the teeth are much alike except for size, and serve only for seizing. A related difference concerns the structure of the jaw itself. In reptiles the jaw is made up of several (usually seven) separate bones. In mammals these are reduced to a single bone for more efficient chewing. The increased activity of mammals also led to improvements in the sensory and nervous systems. The brain increased in size and the senses of smell and hearing became more acute. Reptiles, for example, have only a single "ear bone" for the transmission of sound from the ear drum to the inner ear; in mammals there are three ear bones for more precise hearing. Fossil skulls show that the two additional mammalian ear bones developed from the two bones that form the jaw joint in reptiles.

The shelled reptilian egg was an important advance over the aquatic requirements of the amphibians, yet it had the disadvantage of being easily destroyed by land predators. With one startling exception that we will mention later, mammals have abandoned the external egg. Instead, the egg is retained within the body of the female where the embryo can develop, protected, before being born alive. A few fish and reptiles hold the eggs within the body cavity until the young hatch, but only in mammals has this become an almost universal adaptation. Along with this internal embryonic development, mammals developed another specialization that permits the young animals to become still more fully developed before beginning an independent life—the milk-producing mammary glands from which the class takes its name. The nourishment of milk, and the maternal care that goes with it, help insure that the young mammal will reach maturity without becoming a meal for some predator.

The exact boundary between fossil reptiles and mammals is hard to draw because in Triassic time several separate lines of mammallike reptiles independently developed more and more mammalian jaw and tooth patterns. It is impossible to determine from fossils alone whether or not these animals had hair, mammary glands, and other mammalian characteristics, and thus we cannot be sure that they were not already true mammals. It is most probable, however, that a single line of Triassic insect-eating, mammallike reptiles (Fig. 6-16) gave rise to all the true mammals.

Mesozoic Mammals

The oldest undoubted mammalian fossils are teeth, jaw fragments, and rare skulls of small, shrewlike forms found in Late Triassic rocks (Fig. 6-17). Mammals therefore originated at about the same time as the dinosaurs, but throughout the Mesozoic Era of dinosaur dominance they remained small, inconspicuous animals that are relatively uncommon fossils. Although many mammalian skull features were already present in the ancestral mammallike reptiles, in posture these reptiles were still relatively primitive and inefficient.

FIG. 6-16 Mammallike reptiles. (Above) A well-preserved skull of a small Triassic mammallike reptile of the group which most probably gave rise to the mammals. Scale indicates one inch. (Courtesy A. W. Crompton.) (Below) Restoration showing the probable life appearance of the animal which was about one foot long. (Robert Bakker reconstruction.)

Indeed, the ultimate extinction of the mammallike reptiles and their replacement by the dinosaurs was probably due to their clumsy pattern of limb attachment and locomotion. In spite of their more advanced jaws and feeding habits, mammallike reptiles could not compete successfully with the dinosaurs because they lacked an efficient arrangement of the limbs. Most Mesozoic mammals probably also had this sprawling limb pattern. Although not closely related to modern rodents, they were generally mouselike animals that fed on seeds and insects. The largest was about the size of a house cat (Fig. 6-17). Most were evolutionary dead ends that failed to lead to more advanced groups, but during the Cretaceous Period two groups arose that were to become the ancestors of the many kinds of mammals that dominated the land after the extinction of the dinosaurs. The descendants of these two groups were to become extraordinarily successful because they not only had the warm-bloodedness and increased activity of mammals, but also developed the efficient locomotion patterns of the dinosaurs.

FIG. 6-17 Mesozoic mammals. (Above) Fossil jaw bone of a Jurassic mammal. Mesozoic mammals were small, inconspicuous contemporaries of the dinosaurs. They are known principally from fragmentary fossil remains, such as the jaw bone shown here. The specimen is about one inch long. (Her Majesty's Geological Survey Photographs, Crown Copyright.) (Below) Drawing showing probable life appearance of a Mesozoic mammal. The species shown is among the largest known Mesozoic mammals but is only the size of a house cat. These are illustrated attacking a small reptile, but most of the smaller Mesozoic mammals probably fed on plants and insects. (Neave Parker reconstruction, courtesy illustrated London News.)

FIG. 6-18 Monotremes, egg-laying mammals. These two strange mammals are found living today in Australia. Although they lay eggs in the fashion of reptiles, they have fur and mammary glands for suckling the young after hatching. They have almost no fossil record, but are thought to be direct descendants of the early mammallike reptiles. (Above) Duck-billed platypus. (At right) Spiny anteater. (Both courtesy New York Zoological Society.)

Monotremes

Before discussing the great Cenozoic evolutionary expansion of the mammals, we must say something about two bizarre little animals that are found living today in Australia. These are the *duck-billed platypus* and *spiny anteater,* the only members of the mammalian Order Monotremata (Fig. 6-18). These "monotremes" are warm-blooded, have fur and mammary glands, and in most respects are characteristic mammals. Unlike any other living mammals, however, they do not give birth to living young, but instead *lay eggs* exactly as did their reptilian ancestors. After the eggs hatch, the young are suckled and cared for in the typical mammalian fashion. In addition to this extraordinary pattern of reproduction, monotremes have the inefficient, sprawling limb arrangement found in mammallike reptiles. Unfortunately, monotremes are virtually unknown as fossils, but the egg-laying habit and primitive posture suggest that they are a unique side branch of mammals that descended directly from mammallike reptiles without giving rise to any more advanced groups. Apparently they have survived relatively unchanged since the Mesozoic Era in Australia because, as we shall see, there they were relatively free from competition and predation by more advanced mammals.

MAMMAL DIVERSIFICATION

Following the evolutionary pattern seen so often in other groups, the monotremes and extinct orders of Mesozoic mammals represent an early experimental phase of mammalian history. These were relatively unsuccessful, but two groups that arose from them quickly expanded to dominance after the reptilian extinctions of Late Cretaceous time (Fig. 6-19). These were the marsupial and placental mammals, the groups mentioned earlier as having developed more efficient patterns of locomotion.

Marsupials were the less successful of the two, for they make up only about five percent of all Cenozoic mammals. They differ from the more successful placentals principally in mode of reproduction. In marsupials the young are born alive but at a very early stage of development. These tiny, immature young

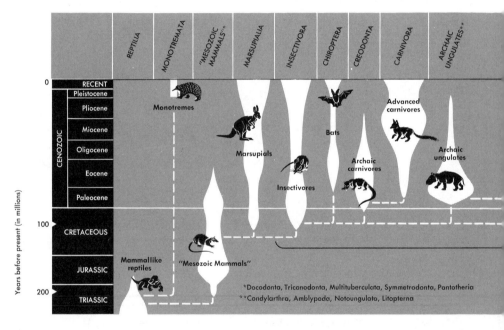

FIG. 6-19 Evolutionary history of the mammals. The dashed lines show the most probable evolutionary relations of the groups. The width of the white areas indicates the approximate abundance of each group.

then crawl into a pouch on the mother's abdomen, attach themselves to a mammary nipple, and complete their development in this external pouch. This pattern is found in all living marsupial mammals (including the familiar pouched kangaroos and oppossums) and was almost certainly characteristic of fossil marsupials as well. In contrast, placental mammals develop inside the uterus of the female, in a special structure, the *placenta*, which allows the developing young to be nourished directly by the mother's body fluids. This structure permits a long period of development within the body of the female, so that birth can be delayed until the young are relatively mature and independent. It is impossible to infer the mode of reproduction directly from fossil skeletons, but marsupial and placental mammals also differ in minor features of skeletal and skull structure, and thus it is usually possible to distinguish fossil skeletons of the two groups.

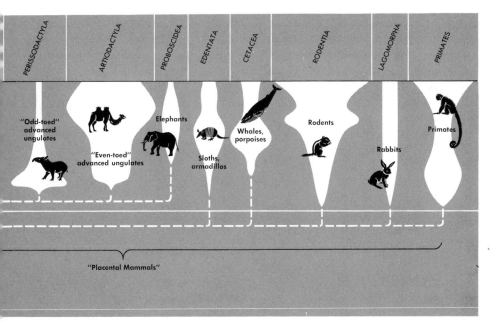

Marsupial Mammals

Although they represent only a single mammalian order and make up only five percent of Cenozoic fossil mammals, marsupials have been of extraordinary evolutionary importance in South America and Australia. Not so, however, in North America, Europe, Asia, and Africa. These continents have been consistently interconnected throughout the Cenozoic Era, and for this reason Cenozoic mammals could easily migrate from one continent to another. These continents, which are often grouped together as the "World Continent," have therefore had rather similar Cenozoic mammalian faunas made up principally of placental forms. In contrast, South America and Australia have been isolated from the other continents through much of Cenozoic time, and have had entirely different mammalian histories. Early in the Cenozoic Era, primitive marsupials reached Australia, but placental mammals, except for flying bats, a few rodents, and forms introduced by man, never became established there. As a result, marsupials in Australia went through a separate Cenozoic evolutionary expansion, which in many ways paralleled that of the placentals on the World Continent. From a primitive, opossumlike ancestor arose several kinds of marsupial herbivores, such as the familiar kangaroos, wallabies, wombats, and koala "bears." In addition, marsupial carnivores evolved that are strikingly similar externally to placental dogs and cats. Mammalian history in South America is not so simple, because the connection to North America was re-established late in the Cenozoic Era, leading to a mixing of primitive and advanced mammals. Unlike the situation in Australia, both primitive marsupials and placentals reached South America early in the Cenozoic Era. Before the later arrival of more advanced placental forms from North America, several groups of strange placental herbivores (including the surviving armadillos and sloths) had developed in South America along with some large, wolflike marsupial carnivores. Most of these groups could not compete successfully with later placental immigrants from North America and became extinct late in the Cenozoic Era.

Placental Mammals

Even though the exotic mammalian faunas of Australia and South America are of great evolutionary interest, they are but minor side issues in the central drama of mammal evolution that took place on the interconnected World Continent. After the extinction of the dinosaurs, the primitive placental stock that appeared in Late Cretaceous time went through a phenomenal evolutionary radiation that led to all the many surviving and extinct orders of placental mammals.

The ancestral placental mammals were small, superficially mouselike representatives of the Order Insectivora. This order includes the modern shrews, moles, and hedgehogs, all of which show primitive anatomical features (Fig. 6-20). In Paleocene and Eocene time insectivores went through an explosive evolutionary expansion that ultimately led to such diverse mammals as whales,

FIG. 6-20 A modern insectivore, the tree shrew, a small forest dweller from Southeast Asia. Similar shrewlike insectivores are believed to be the ancestral stock which gave rise to all other placental mammals. (Courtesy New York Zoological Society.)

bats, horses, and elephants. The earliest representatives of most orders of placental mammals did not, however, closely resemble their modern counterparts, but were generally smaller and less specialized. For example, ancestral catlike carnivores were no larger than a rabbit, and horses were four-toed animals about the size of a fox terrier (Fig. 6-21). Throughout the Oligocene and Miocene Epochs, some members of each order tended to increase in size and to diversify into increasingly specialized types. Some orders, such as the Perissodactyla (horses and their relatives) and the Proboscidea (elephants), reached their evolutionary climax earlier in Cenozoic time and are reduced in numbers and diversity today. Other orders, such as the Rodentia and Primates, appear to have steadily increased in diversity up until modern times.

Just as with the Mesozoic reptiles before them, mammals have not only become the dominant herbivores and carnivores of the land surface, but have also developed flying forms—the bats—and have returned to the sea as marine carnivores—whales, porpoises, seals. The principal large mammal carnivores of the land are dogs, cats, bears, and their relatives, all of which are included in the Order Carnivora. The greatest diversification of land mammals has involved the many herbivorous orders; these range in size and habit from tiny rodents, to tree-living monkeys, to huge elephants, giraffes, and rhinoceroses.

In addition to the familiar surviving groups of mammals, there are several important extinct orders. Most of these were "archaic ungulates," hoof-bearing herbivores that superficially resembled a large dog or sheep. Some, however, were as large as a modern rhinoceros. Several groups of these primitive ungulates arose from the insectivores early in the Cenozoic Era and were the dominant larger land herbivores during the Paleocene and Eocene Epochs. In Late Eocene time they were beginning to be replaced by the modern hoof-bearing orders (Artiodactyla and Perissodactyla), and most were extinct by the beginning of the Oligocene Epoch. Along with these archaic ungulates, an early group of carnivores—the creodonts—evolved that were later to give rise to the modern Order Carnivora.

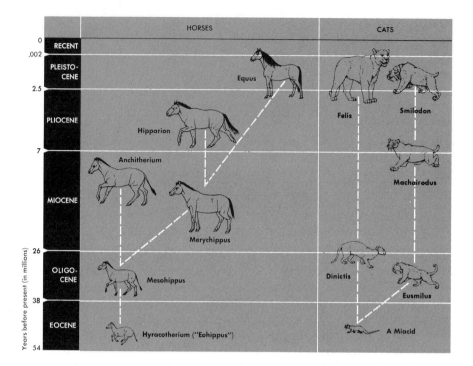

FIG. 6-21 Evolutionary history of horses and large carnivorous cats.

Aside from the creodonts and primitive ungulate orders, few orders of placental mammals have become extinct; the general history of placental mammals since mid-Cenozoic time has therefore been a progressive modernization leading to the familiar mammals of today. Recently, however, there have been some dramatic extinctions of a few large species in several orders. Until only a few thousand years ago, huge relatives of the elephant (mammoths and mastodons), large sabre-toothed cats, and giant armadillos and ground sloths still lived in North America. All of these died out rather suddenly, possibly as a result of hunting by early man.

seven

man

We have now completed our survey of the history of life except for the final and most significant evolutionary event of all—the origin of our own species, *Homo sapiens*. The broad outlines of human evolution have been understood for many years, but within the last decade important new fossil discoveries, and reinterpretation of previously discovered specimens, have greatly clarified the origin of man and the history of his primate ancestors. This work is still continuing at a rapid pace and the whole subject of human evolution is today among the most active and exciting areas in the history of life.

In discussing human evolution, it will be necessary to retrace our steps a bit by considering in more detail the adaptations and history of man's closest relatives—the "premonkeys," monkeys, and apes—that, along with man and his extinct ancestors, make up the mammalian Order Primates. We shall see that the development of the Hominidae, man's own family of large-brained, tool-using primates, is an extremely recent evolutionary development that has mostly taken place within the last few million years (Fig. 7-1). Nevertheless, we shall find that the characteristics that led to man's rapid ascendance were being established and refined by his primate ancestors throughout the preceding 70 million years that make up the Cenozoic Era.

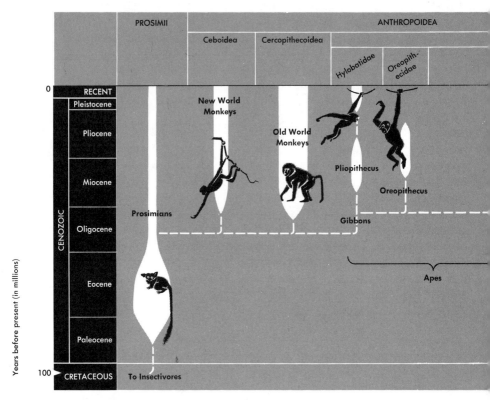

FIG. 7-1 Evolutionary history of the primates. The dashed lines show the most probable evolutionary relations of the group. The width of the white areas indicates the approximate abundance of each group.

PRIMATE ORIGINS AND ADAPTATIONS

In the last chapter we saw that most of the orders of placental mammals arose early in the Cenozoic Era from simple, shrewlike ancestors of the Order Insectivora. The Order Primates is no exception, for the oldest fossil remains of primates—isolated teeth and jaws from Early Paleocene rocks—are almost indistinguishable from insectivores. Most insectivores live on the ground and, as the name implies, are principally insect-eaters. Primates diverged from the insectivores by becoming adapted to a different mode of life, that of the omnivorous tree-dweller. This new habit led to fundamental changes in the skeletal pattern that were to be of great importance in the later evolution of the primates.

FIG. 7-2 (Facing page) Two unique primate adaptations, the grasping hand and stereoscopic vision, both of which originated as specializations for life in trees. The primitive insectivore's fan-shaped arrangements of digits (left) can only grab by digging in with claws, but the primate's opposable thumb allows a precise grip. The overlap of right and left eye vision is narrow in the insectivore and broad in the primate because of the different placement of eyes in the head.

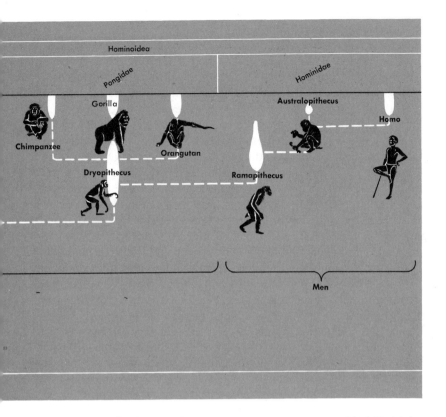

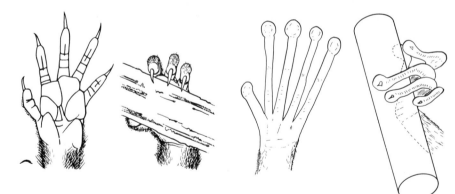

Insectivore   Primate (Prosimian)

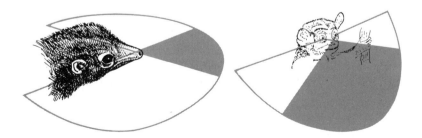

The most significant of these changes was the development of the grasping hand with its opposable thumb (Fig. 7-2), an almost universal primate characteristic that originated as a means of grasping limbs and branches for moving about in trees. Because of their grasping hand, primates are the most successful and efficient tree-dwelling mammals; other mammal groups have developed tree-dwellers (such as squirrels, which are rodents), but all of these depend on claws or other less efficient means of holding on. The development of the primate hand was accompanied by another specialization of far-reaching importance. In most mammals, the eyes are placed toward the sides of the head in a position that allows a maximum field of vision from each eye, but prohibits precise stereoscopic vision because there is little overlap in the visual fields. In primates, stereoscopic vision was necessary for confident movement high in the tops of trees where a slight misjudgment in depth perception could lead to a fatal fall. Correlated with the grasping hand and the tree-dwelling habit, then, were a forward migration of the eyes and a relative flattening of the face, so that the eyes could stereoscopically view the same area (Fig. 7-2). The grasping hand and stereoscopic vision arose in primates as adaptations for life in trees, but they were to become the foundation of man's later evolutionary success on the ground, for they permitted him to grasp tools and weapons, and to coordinate hand and vision precisely in such activities as hunting and tool-making.

Prosimians

The grasping hand and stereoscopic vision originated early in primate history; fossil skeletons show that they were developed by Eocene time when small tree-dwelling primates, about the size of a large cat or small monkey, were common in North America and Eurasia. These early primates were not true monkeys but a less advanced group that retained primitive insectivorelike skeletal features. For this reason they are assigned to the Suborder Prosimii ("premonkey"), which includes ancestral primates (Fig. 7-3) that gave rise to the second suborder, the Anthropoidea (monkeys, apes, and men).

Prosimians were widespread and common early in the Cenozoic Era, but decreased dramatically in importance after the origin of true monkeys and apes in Oligocene time. Fortunately, the group did not become extinct but survived to the present day in certain isolated or unusually favorable environments. The most primitive surviving prosimians are the *lemurs*, a group not unlike the Early Cenozoic prosimians. Lemurs are found today only on the island of Madagascar (Fig. 7-4(A)). Madagascar has been separated from Africa since at least Early Cenozoic time, and has developed few mammalian predators. Apparently some primitive prosimians reached the island early in the Cenozoic Era and have developed there free from both predators and competition with more advanced primates. Although these Madagascar prosimians did not give rise to higher forms, they have diversified into many species with differing habits and adaptations. In addition to the lemurs of Madagascar, two other groups of somewhat

FIG. 7-3 An early fossil prosimian (*Smilodectes*) from Eocene rocks. Prosimians are the earliest primates and are the ancestors of the more advanced monkeys, apes, and men. Note the grasping hand and forward position of the eyes. (Above) Mounted fossil skeleton, about 18 inches high. (Courtesy David Linton.) (Below) Drawing showing the probable life appearance of the skeleton. (Maternes reconstruction, courtesy David Linton.)

more advanced prosimians survive today; the large-eyed tarsiers of Borneo, Sumatra, and the Philippines, which have both prosimian and monkeylike characteristics, and the lorises, a lemurlike group found in Africa, India, and Southeast Asia, but not in Madagascar (Figs. 7-4(B) and 7-4(C)).

FIG. 7-4 Present-day examples of prosimians. During the last half of the Cenozoic Era, prosimians have been overshadowed by more advanced monkeys, but three groups, lemurs, tarsiers, and lorises, still survive in parts of Africa and Asia. (A) A lemur from the island of Madagascar. (Courtesy New York Zoological Society.) (B) A tarsier from the Philippine Islands. (Lilo Hess photo, Three Lions, Inc.) (C) A loris from Ceylon. (Courtesy New York Zoological Society.)

PRE-PLEISTOCENE MONKEYS AND APES

The higher primates of the Suborder Anthropoidea include three distinctive groups that are commonly ranked as superfamilies: the Ceboidea, or "New World monkeys," the Cercopithecoidea, or "Old World monkeys," and the Hominoidea, which includes man and the apes. The Old World monkeys of Africa and Asia show many fundamental differences from the superficially similar New World monkeys native to South America. Apparently, the two groups originated independently from prosimian ancestors and then showed parallel evolutionary trends; there is no fossil evidence of migration or genetic interchange between them. Today the third superfamily, the Hominoidea, includes only man and four surviving kinds of apes (chimpanzee, gorilla, orangutan, and gibbon), although there are several extinct representatives of this group. Hominoids differ from monkeys in lacking a tail and in other fundamental adaptive features. Before the relatively recent migrations of man to the New World, hominoids were confined to Africa and Eurasia, where the superfamily apparently originated.

The crucial time of differentiation of all three anthropoid superfamilies was the Oligocene Epoch. In the preceding Paleocene and Eocene Epochs the only fossil primates are prosimians; Miocene deposits yield clearly differentiated and

FIG. 7-5 Primate fossil quarry in the Oligocene Fayum deposits of the Egyptian desert. Oligocene primates are extremely rare, but are unusually significant because the transition from simple prosimians to advanced monkeys and apes took place in Oligocene time. Most known specimens have been collected from these Fayum deposits since 1960. (Left) The fossils are found by carefully brushing away the loose sediment with brushes. (Right) A typical small jaw being uncovered. (Courtesy E. L. Simons.)

essentially modern monkeys and apes. Thus, the prosimian-anthropoid transition must have taken place during Oligocene time. Unfortunately, Oligocene fossil primates are extremely rare and, until very recently, were known only from a half-dozen bones and teeth discovered in the early years of this century in the Fayum area of northern Egypt. Beginning in 1960, a program of modern restudy of the Fayum deposits has yielded over 100 primate specimens, including jaws, limb bones, and skull fragments (Fig. 7-5). Most of these are of Late Oligocene age. Surprisingly, none are prosimians but are instead already clearly differentiated apes or relatives of the Old World monkeys, even though they are structurally primitive and show evidence of their prosimian ancestry. These specimens therefore indicate that the separation of apelike hominoids from monkeys was an early and fundamental event in primate history, one that had already taken place by the close of the Oligocene Epoch.

Monkeys

Fossil primates are more abundant in Miocene and Pliocene deposits than in the transitional Oligocene rocks, yet even in these younger strata, they are among the rarer fossil vertebrates because of their dominantly tree-dwelling habit. We have already noted that the New World monkeys evolved apart from the apes and monkeys of the Old World. The first fossil New World monkeys are found in Late Oligocene and Early Miocene deposits of South America and are very similar to their modern descendants. All earlier fossil primates from the Western Hemisphere are prosimians, and thus the fossil record throws little light on the prosimian-monkey transition in the New World. Aside from the transitional Fayum Oligocene specimens, fossil Old World monkeys also closely resemble their modern relatives.

Apes

The fossil record of apes is more complete than that of monkeys, a fortunate circumstance since our own ancestry lies in this primate group. Miocene and Pliocene deposits of Europe, Africa, and Asia have yielded several hundred specimens, mostly teeth and jaw fragments, of a group of generalized apes called *dryopithecines* (Fig. 7-6). These early apes appear to have been ancestral both to man and to the modern chimpanzee, gorilla, and orangutan. The dryopithecines show the same size range as do modern apes; some were about the size of a small gibbon, others were as large as a gorilla. All, however, were alike in dental pattern, a similarity suggesting that the different sizes only represent widely distributed species or individuals of a single genus (formerly often called *Proconsul*, but now more correctly termed *Dryopithecus*). Although no complete skulls or skeletons of dryopithecines are known, partial skulls and jaws show that they had a relatively small, monkeylike brain, but were apelike or even manlike in the pattern of jaws, face, and dentition. Unfortunately, only a few fragmentary limb bones have as yet been discovered, so it is uncertain whether these animals were still principally tree-dwellers, or had become adapted to life

FIG. 7-6 Dryopithecines, generalized fossil apes from Miocene and Pliocene rocks. These fossil apes were probably ancestral to both men and present-day apes. (Top) Two views of a well-preserved fossil skull found in Africa in 1948. The skull is about four inches high. (Right) Drawing showing the probable life appearance of a dryopithecine. Only a few fragmentary fossil limb bones have so far been discovered, and the upright posture is still conjectural. (By permission of the Trustees of the British Museum (Natural History), Wilson reconstruction.)

on the ground. Tree-dwelling, as we have noted, is the basic primate adaptation found in all prosimians and in most monkeys, although some Old World monkeys—for example, the baboons—have secondarily adapted to life on the ground. Modern hominoids include both tree-dwellers (gibbon and orangutan) and forms that spend much or all of their lives on the ground (chimpanzee, gorilla, and man). The large size of certain dryopithecine specimens suggests that some may have also been adapted for life on the ground, but the evidence is still too scanty to be certain.

Besides the widely distributed, relatively unspecialized dryopithecines, there are other less common and more specialized Miocene and Pliocene fossil apes. One such form from Europe and Africa, called *Pliopithecus*, is probably off the main line of human evolution, but may have given rise to the gibbons. Another genus, *Oreopithecus*, is known from many fragments and one nearly complete skeleton from coal-bearing rocks of Italy dating from Miocene and Pliocene times. It was a long-armed tree-dweller that apparently became extinct without giving rise to other forms.

There is one final genus of Miocene-Pliocene fossil hominoid that, although known only from a half-dozen jaw fragments found in India and Africa, is of great significance because the pattern of dentition, unlike that of modern apes, is extraordinarily manlike (Fig. 7-7). This genus, called *Ramapithecus*, probably arose from a dryopithecine ancestor early in Miocene time and is apparently a direct ancestor of modern man.

FIG. 7-7 *Ramapithecus*, the oldest fossil man, from Miocene and Pliocene deposits of India and Africa. The genus is known only from a few teeth and skull fragments, but these are extraordinarily manlike. The photographs show two views of one of the larger fragments preserving four teeth. The specimen was collected in India in 1932. (Courtesy Peabody Museum of Natural History, Yale University.)

THE PLEISTOCENE EXPANSION OF MAN

Several million years ago, subtle and as yet poorly understood climatic and topographic changes led to the first of a series of great continental glaciers that were, at times of maximum expansion, to cover large areas of the Northern Hemisphere with a thick sheet of ice. This interval of expanding and contracting continental ice sheets is the Pleistocene Epoch, which is continuing today inasmuch as the last great ice cap contracted only a few thousand years ago and still covers most of Greenland. Such periods of widespread continental glaciations are apparently rare events in Earth history; the only comparable glaciations for which there is convincing geologic evidence took place in Permian and Precambrian times. The Pleistocene Epoch thus probably represents the first time of widespread glaciations since before the origin of mammals. Surprisingly, the extreme climatic changes of Pleistocene time appear to have led to relatively few extinctions of animals and plants. Although a number of large land mammals, such as mammoths, mastodons, ground sloths, and sabre-tooth cats, became extinct late in Pleistocene time, these extinctions appear to have taken place so recently that they are probably a result of hunting by early man rather than of climatic change. The principal effect of the advancing and receding Pleistocene ice sheets has been to cause rapid changes of distribution in animal and plant species. This, then, is the rather exceptional geologic setting in which our own species, *Homo sapiens*, has developed. The expansion of man in the last few million years is a rapid and recent evolutionary event when compared with the 600 million years that have passed since animal life first became abundant in Early Cambrian time. On the other hand, when we remember that the oldest fragments of written human history go back only about 5,000 years, we can appreciate that the several million years of the "prehistoric" human record are a long time indeed.

As noted earlier, the evolutionary line leading to man passes through Miocene and Pliocene dryopithecines and *Ramapithecus*, both of which are known mostly from fossil jaws and teeth. It is still uncertain when man's ground-living habit and characteristic upright posture originated, for these habits can be definitely established only from well-preserved limb or pelvic bones. Most probably, *Ramapithecus* already was a bipedal ground-dweller, for its teeth suggest a diet, and thus life habits similar to modern man's. In any event, fossil limb bones show that man's upright posture and ground-dwelling habit were already established in the earliest Pleistocene descendants of *Ramaphithecus*.

At this point we must digress for a moment and consider just what it is that makes man different from his primate relatives. We have seen that man's grasping hand and precise vision are universal primate features. Other primates, such as baboons, chimpanzees, and gorillas, spend most of their lives on the ground and can even walk in an upright position, although not so effectively or consistently as we do. The use of tools is commonly considered to be a distinguishing characteristic of man, but even this habit is occasionally found in other primates. Chimpanzees, for example, fashion short sticks from twigs, which they then thrust into ant or termite holes, withdraw, and lick off a meal. Man's *most* characteristic feature, in which he differs from all of his primate relatives, is his extraordinarily large brain and high intelligence; thus the Pleistocene history of man is primarily a story of progressive enlargement of the brain. Interrelated was an increasing use of tools, culminating in the development of agriculture and the domestication of animals about 7,000 years ago.

Fortunately, the Pleistocene evolutionary sequence of man is more clearly revealed in the fossil record than is that of his pre-Pleistocene ancestors. Until very recently, the record of Pleistocene man was thought to include many species and genera, because in the past each new fossil specimen was generally given

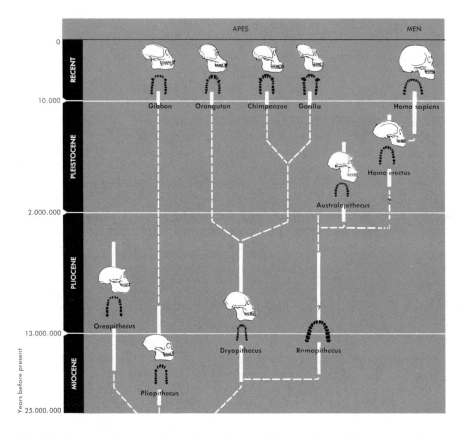

FIG. 7-8 Summary of the evolutionary history of man and the apes. The solid lines show the known fossil record and the dashed lines the most probable evolutionary relations of the groups. A typical dental pattern is shown below each skull.

a new name, without concern for its similarities to previously discovered specimens. Since the mid-1960's, however, reconsideration of these Pleistocene fossils in the light of modern biological principles has revealed that the story is much simpler than had been formerly believed. In essence, there are three principal stages in man's Pleistocene evolution: *Australopithecus* and primitive *Homo* in the Early Pleistocene, *Homo erectus* in the Middle Pleistocene, and *Homo sapiens* in the Late Pleistocene (Fig. 7-8).

Australopithecus

The oldest well-understood Pleistocene men are now assigned to the genus *Australopithecus* (Fig. 7-9), which is one of only three genera now recognized in man's family, the Hominidae (the other two are *Ramapithecus* and *Homo*). Australopithecines were first discovered in South Africa in 1924 when a single well-preserved skull was found. Since then, many additional skulls, jaws, and limb bones have been collected in South Africa and, more recently, in the Olduvai Gorge of Tanzania. Fragmentary, but probably identical, specimens have also been discovered as far away as Java, suggesting that australopithecines were widely distributed in the Old World. These specimens, all of which appear to be older than 500,000 years, probably represent two species that were generally similar but differed in size. The larger, *Australopithecus robustus*, was about the size of a gorilla; the smaller, *Australopithecus africanus*, was chimpanzee-sized. Both species were ground-dwellers with upright posture, for their limb bones are almost identical with those of modern man. The brain volume was, however, only about 600–700 cubic centimeters, or about half that of modern man. In shape if not size, the skulls, like the limb bones, are remarkably manlike. In short, australopithecines were men in every feature except their smaller brains.

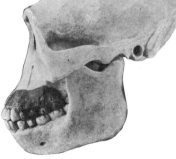

FIG. 7-9 *Australopithecus*, Early Pleistocene men, known from many fossil skulls and skeletal bones from Africa, and a few fragments from Asia and the Middle East. The skeletons have the upright posture of modern man, but differ primarily in having a brain only half the size of present-day men. (Above) The probable life appearance of *Australopithecus*. (From J. Augusta and Z. Burian, *Prehistoric Man*, Spring Books, London.) (At right) Reconstructed skull based on several relatively complete specimens. The skull is about 11 inches high. (Courtesy American Museum of Natural History.)

The evolution of man's upright posture therefore preceded, and probably paved the way for, later expansion in brain size. The upright stance completely freed the hands from use in locomotion; instead they could now be used exclusively for such tasks as toolmaking and weapon-throwing. These tasks, in turn, put a premium on increased intelligence and awareness and thus led to selection pressures for a larger brain. In this regard it is most significant that several sites bearing australopithecine fossils also contain crude stone tools as well as the fractured skulls of baboons and other animals, indicating that man's Early Pleistocene relatives already were toolmakers and weapon-users in spite of their relatively small brains.

HOMO

Until rather recently, all Early Pleistocene men were thought to be australopithecines, which are succeeded in Middle Pleistocene time by a widely distributed and larger brained species, *Homo erectus*, that is so like modern man that it clearly belongs in the same genus. In 1973, however, an Early Pleistocene skull was found in northern Kenya that is far more manlike than its australopithecine contemporaries. Since then, additional discoveries in East Africa have shown that *Homo*-like fossils clearly occur in far earlier Pleistocene deposits than had been previously suspected. These discoveries suggest that *Homo* may not be a direct descendant of the abundant Early Pleistocene australopithecines; instead, it appears more probable that both *Homo* and advanced *Australopithecus* diverged from a common ancestor in Late Pliocene time. It is still likely, however, that this ancestral form more closely resembled *Australopithecus* than the larger-brained *Homo*. Whatever the exact time of origin of the genus *Homo*, the fossil record makes it clear that by Middle Pleistocene time the australopithecines were becoming extinct, while a single species of man, *Homo erectus*, was becoming widely distributed throughout the Old World.

Homo erectus first became known when a skull cap was found in Java in 1891; this was one of the first human ancestors discovered. Thirty years after the initial discovery of this "Java Man," additional specimens were found in China, and since that time many additional skulls, jaws, and limb bones have been reported from Europe and Africa as well as in Asia. As with *Australopithecus*, these specimens have been given many different names, but modern study shows that they are best interpreted as a single species that differed from present-day man principally in having a brain volume of 900–1,100 cubic centimeters, which is about intermediate between that of *Australopithecus* (600–700 cubic centimeters) and modern man (1,400–1,600 cubic centimeters) (Fig. 7-10). Like *Australopithecus*, *Homo erectus* used stone tools, principally large hand axes made from pebbles of flint that were sharpened on one side by chipping. These tools show a progressive advance in design and workmanship through Middle Pleistocene time, indicating that man was becoming a more adept toolmaker as his brain size increased.

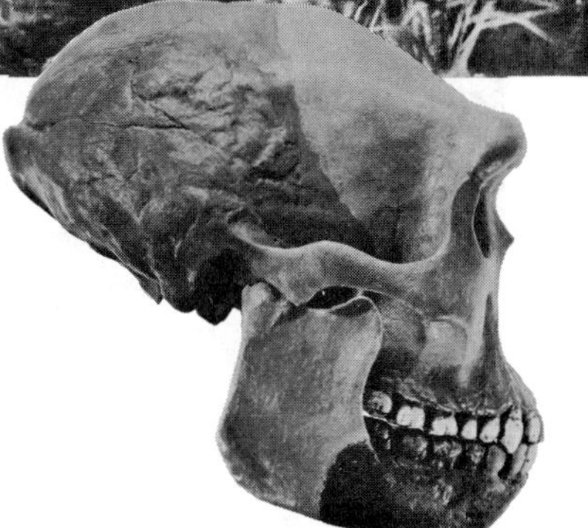

FIG. 7-10 *Homo erectus*, Middle Pleistocene men, known from many fossil skulls and skeletal bones from Asia, Africa, and Europe. The skeletons are like modern men, but have a brain intermediate in size between that of *Australopithecus* and *Homo sapiens*. (Above) The probable life appearance of *Homo erectus*. (From J. Augusta and Z. Burian, Prehistoric Man, Spring Books, London.) (At right) Reconstructed skull based on several relatively complete specimens. (Weidenreich restoration, courtesy American Museum of Natural History.)

Most specimens of *Homo erectus* have been found in sediments that range in age from about 700,000 down to about 200,000 years. This species is believed to be the direct ancestor of our own species, *Homo sapiens*, which first occurs in the fossil record about 500,000 years ago, and thus was contemporaneous with *Homo erectus* for about 200,000 years.

As with *Australopithecus* and *Homo erectus*, fossil specimens of early *Homo sapiens* commonly have been described as other species. The best-known example is "Neanderthal Man," a large-boned race that lived about 100,000 years ago (Fig. 7-11). These fossils were formerly assigned to a separate species, *Homo neanderthalensis*, but it is now evident that they are too similar to modern man to justify this separation.

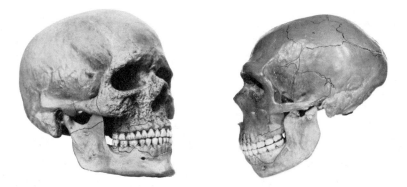

FIG. 7-11 Fossil *Homo sapiens*, Late Pleistocene men. (Left) Reconstructed skull of "Cro-Magnon" man, an early race similar to modern man. (Courtesy Field Museum of Natural History.) (Right) Reconstructed skull of "Neanderthal" man, a heavy-boned race of *Homo sapiens* that lived in Europe and Asia about 100,000 years ago. (Courtesy American Museum of Natural History.)

FIG. 7-12 The development of toolmaking cultures by Pleistocene man.

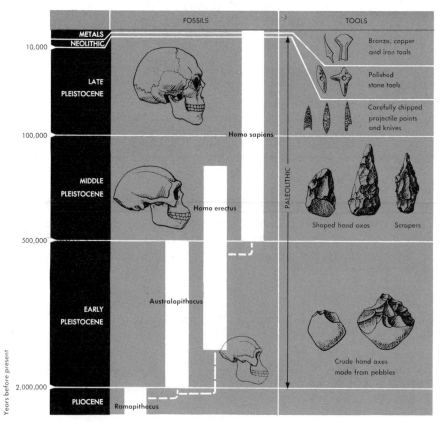

THE EVOLUTION OF HUMAN CULTURE

With the origin of *Homo sapiens* we have reached the boundary between the natural sciences of biology and geology and the social sciences of archeology, cultural anthropology, and ancient history. So far we have stressed only the direct fossil evidence for man's ancestry, but throughout the Pleistocene Epoch fossil skulls, bones, and teeth of men are extremely rare in comparison with his stone tools, which of course are much more resistant to decay. Study of these tools by archeologists has established a sequence of cultural events that can, in a general way, be correlated with man's physical evolution. To discuss the details of this cultural evolution would require another book at least as long as this one, but a broad outline will be an appropriate final topic for our survey of the history of life.

Anthropologists recognize three principal stages in man's cultural evolution: the Paleolithic, or "Old Stone Age," characterized by chipped stone tools; the Neolithic, or "New Stone Age," characterized by ground and polished stone tools; and, finally, the Age of Metals when tools were made first of copper and bronze, later of iron (Fig. 7-12). Most of man's Pleistocene history falls into the Paleolithic "Old Stone Age," for Neolithic cultures began only about 10,000 years ago, and were followed about 5,000 years ago by metal-using cultures. Thus, all of the cultural history of *Australopithecus* and *Homo erectus*, as well as most of the existence of *Homo sapiens*, is represented by Paleolithic chipped stone tools. As we have noted, Paleolithic tools show a progressive improvement in toolmaking techniques throughout the Pleistocene Epoch. The earliest are crude hand axes, and the most advanced are carefully styled projectile points and knives made by Late Paleolithic *Homo sapiens* (Fig. 7-12). Late Paleolithic man also used bone for making fine tools, such as needles, and had a highly developed artistic sense, as is evident from elaborately carved bone objects and from the well-known Late Paleolithic cave paintings of France and Spain (Fig. 7-13). Between 10 and 15 thousand years ago, the chipped stone tools of Late Paleolithic time gave way to more advanced stone tools made by grinding and polishing. The Neolithic peoples who made these tools developed still more important cultural advances; they learned to make pottery for storage of food and water and, most important of all, they began to cultivate plants and domesticate animals. With this step, for the first time in his long Pleistocene history, man no longer had to depend on hunting and the gathering of wild plants, but could grow his own food. The development of agriculture by Neolithic man was one of the most significant events in all of human history, for it not only permitted the development of permanent communities, some of which were later to become the first cities, but also allowed a division of labor—some men provided the food, while others became craftsmen, priests, tradesmen, and scholars. After the development of agriculture in this "Neolithic Revolution," human cultures evolved rapidly. About 5,000 years ago, metals were first used

for tools, and, at about the same time, the development of writing in Egypt and Mesopotamia led to the beginnings of recorded human history.

The Pleistocene record of fossil man and his tools still leaves unanswered many fundamental questions of cultural evolution. We have, for example, only vague hints about such crucial events as the origin of language, the development of clothing, and the social structure of early human communities. In spite of these areas of ignorance, man's development from his Cenozoic primate ancestors is now among the better-understood events in the history of life.

appendix:
a classification of organisms

This classification includes all the generally recognized phyla of organisms. Symbols shown in the key indicate the dominant habitat (marine or terrestrial or both) of each group. Phyla listed for completeness but not discussed in the text are indicated by gray symbols. All the subphyla of the Phylum Tracheophyta and Phylum Chordata are listed, as are all the classes of seed plants and vertebrate animals. Many minor classes of the larger phyla of invertebrate animals (Mollusca, Arthropoda, Echinodermata) are omitted, as are a few minor orders of reptiles and mammals. Extinct groups are shown in italics. Boldface figures indicate the pages on which groups are illustrated.

Key: ◆ Dominantly marine
■ Dominantly terrestrial, land or fresh water (▨ not discussed in text)
● Both marine and terrestrial (◉ not discussed in text)
Italic Extinct groups

PROCARYOTES

● PHYLUM SCHIZOMYCOPHYTA: bacteria **17, 18**
● PHYLUM CYANOPHYTA: blue-green algae **14, 17, 18**

EUCARYOTES—PLANTS

▨ PHYLA MYXOMYCOPHYTA: slime molds,
EUGLENOPHYTA: plantlike flagellates
◆ PHYLUM PYRROPHYTA: dinoflagellates **45**
● PHYLUM CHLOROPHYTA: green algae
■ PHYLUM CHAROPHYTA: stoneworts
◆ PHYLUM PHAEOPHYTA: brown algae **45**
◆ PHYLUM RHODOPHYTA: red algae **45**
● PHYLUM CHRYSOPHYTA: diatoms, coccolithophorids,
golden-brown algae **45, 48**

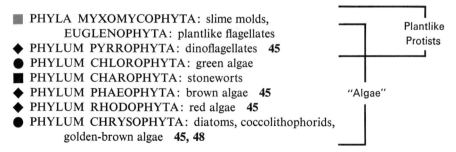

Plantlike Protists

"Algae"

155

● PHYLUM MYCOPHYTA: fungi
■ PHYLUM BRYOPHYTA: mosses, liverworts   **71**
■ PHYLUM TRACHEOPHYTA: vascular plants
   ■ SUBPHYLUM PSILOPSIDA: psilopsids   **95**
   ■ SUBPHYLUM LYCOPSIDA: lycopsids   **96, 99**
   ■ SUBPHYLUM SPHENOPSIDA: sphenopsids   **97, 99**
   ■ SUBPHYLUM PTEROPSIDA: ferns   **98**
   ■ SUBPHYLUM SPERMOPSIDA: seed plants
     ■ *Class Pteridospermophyta: seed ferns*   **91, 99, 102**
     ■ CLASS CYCADOPHYTA: cycads   **92, 103**
     ■ CLASS GINKGOPHYTA: ginkgoes   **92, 104**
     ■ CLASS CONIFEROPHYTA: conifers   **93**
     ■ CLASS ANGIOSPERMOPHYTA: flowering plants,
       or "angiosperms"   **105, 106**

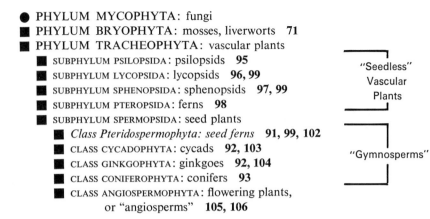

"Seedless" Vascular Plants

"Gymnosperms"

EUCARYOTES—ANIMALS

● PHYLUM SARCODINA: amebas, foraminiferans,
   radiolarians   **45, 48, 58**
● PHYLA MASTIGOPHORA: animallike flagellates,
   SPOROZOA: sporozoans, CILIOPHORA: ciliates
◆ PHYLUM PORIFERA: sponges   **45, 58**
◆ PHYLUM COELENTERATA: jellyfishes, corals   **23, 24, 45, 58**
● PHYLA CTENOPHORA: comb jellies, PLATYHEL-
   MINTHES: flatworms, MESOZOA: mesozoans,
   RHYNCHOCOELA: ribbon worms, ASCHEL-
   MINTHES: aschelminths, ACANTHOCEPH-
   ALA: spiny-headed worms, ENTOPROCTA:
   entoprocts, PHORONIDA: phoronids, CHAETO-
   GNATHA: arrow worms, POGONOPHORA:
   pogonophorans
◆ PHYLUM BRYOZOA: bryozoans   **50, 58**
◆ PHYLUM BRACHIOPODA: brachiopods   **25, 45, 50, 55, 58**
● PHYLUM MOLLUSCA: molluscs
   ● CLASS GASTROPODA: snails   **50, 58**
   ● CLASS BIVALVIA: clams   **45, 58**
   ◆ CLASS CEPHALOPODA: cephalopods   **45, 59**
● PHYLUM ANNELIDA: segmented worms   **23, 24, 45, 59**
● PHYLUM ARTHROPODA: arthropods
   ◆ *Class Trilobita: trilobites*   **25, 59**
   ● CLASS CRUSTACEA: crustaceans   **45, 50, 59**
   ■ CLASS ARACHNIDA: arachnids   **73**
   ■ CLASS INSECTA: insects   **73**
◆ PHYLUM ECHINODERMATA: echinoderms
   ◆ CLASS CRINOIDEA: crinoids   **50, 59**
   ◆ CLASS ASTEROIDEA: starfish   **50, 59**
   ◆ CLASS OPHIUROIDEA: serpent stars   **50, 59**
   ◆ CLASS ECHINOIDEA: sea urchins   **45, 59**

Animallike Protists, or "Protozoans"

"Invertebrates"

■ *Problematic extinct invertebrate groups: Archeocyatha,*
     Stromatoporoidea, Graptolithina

● PHYLUM CHORDATA: chordates

  ◆ SUBPHYLUM HEMICHORDATA: acorn worms **76**   "Invertebrates"

  ◆ SUBPHYLUM UROCHORDATA: tunicates **76**

  ◆ SUBPHYLUM CEPHALOCHORDATA: lancelets **45**

  ● SUBPHYLUM VERTEBRATA: vertebrates

    ● CLASS AGNATHA: jawless fishes **78, 80**

    ● *Class Placodermi: placoderms* **78, 82**

    ◆ CLASS CHONDRICHTHYES: sharks, rays **78**

    ● CLASS OSTEICHTHYES: bony fishes

      ● SUBCLASS ACTINOPTERYGII: ray-finned fishes   "Fishes"
        45, 78, 83

      ● SUBCLASS SARCOPTERYGII: lobe-finned fishes **79, 83**

        ■ ORDER DIPNOI: lungfishes **79**

        ● ORDER CROSSOPTERYGII: crossopterygians **79, 84, 86**

  ■ CLASS AMPHIBIA: amphibians

    ■ *Subclass Labyrinthodontia: labyrinthodonts* **79, 84, 87**

    ■ SUBCLASS LISSAMPHIBIA: frogs, toads, salamanders **79**

  ■ CLASS REPTILIA: reptiles

    ■ *Order Cotylosauria: stem reptiles* **112**

    ■ *Orders Pelycosauria, Therapsida: mammallike reptiles* **113, 119,**
      129, 132

    ● ORDER CHELONIA: turtles **45, 114**

    ◆ *Order Sauropterygia: plesiosaurs* **114**

    ◆ *Order Ichthyosauria: ichthyosaurs* **114**

    ■ ORDER SQUAMATA: lizards and snakes **114**

    ■ *Order Thecodontia: thecodonts* **113**

    ■ ORDER CROCODILIA: crocodiles **116**

    ■ *Order Pterosauria: pterosaurs* **116, 118, 119**   "Ruling

    ■ *Order Saurischia: theropod and sauropod dinosaurs*   Reptiles"
      120, 124, 125

    ■ *Order Ornithischia: stegosaur, ankylosaur, ornithopod,*
      and horned dinosaurs **122, 123, 124**

  ■ CLASS AVES: birds **127**

  ■ CLASS MAMMALIA: mammals

    ■ ORDER MONOTREMATA: monotremes **131, 132**

    ■ *Orders Docodonta, Triconodonta, Multituberculata,*
      Symmetrodonta, Pantotheria: "Mesozoic mammals" **132**

    ■ ORDER MARSUPIALIA: marsupials **132**

    ■ ORDER INSECTIVORA: insectivores **132, 135, 138, 139**

    ■ ORDER CHIROPTERA: bats **132**

    ■ *Order Creodonta: archaic carnivores* **132**

    ● ORDER CARNIVORA: advanced carnivores—dogs, cats,
      bears, seals **132, 136**   "Placental

    ■ *Orders Condylarthra, Amblypoda, Notoungulata,*   Mammals"
      Litopterna: archaic ungulates **132**

    ■ ORDER PERISSODACTYLA: "odd-toed" advanced
      ungulates—horses, rhinoceroses, tapirs **133, 136**

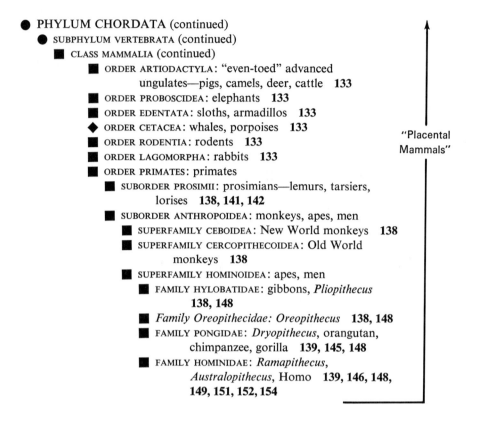

● PHYLUM CHORDATA (continued)
  ● SUBPHYLUM VERTEBRATA (continued)
    ■ CLASS MAMMALIA (continued)
      ■ ORDER ARTIODACTYLA: "even-toed" advanced
        ungulates—pigs, camels, deer, cattle   **133**
      ■ ORDER PROBOSCIDEA: elephants   **133**
      ■ ORDER EDENTATA: sloths, armadillos   **133**
      ◆ ORDER CETACEA: whales, porpoises   **133**
      ■ ORDER RODENTIA: rodents   **133**
      ■ ORDER LAGOMORPHA: rabbits   **133**
      ■ ORDER PRIMATES: primates
        ■ SUBORDER PROSIMII: prosimians—lemurs, tarsiers,
          lorises   **138, 141, 142**
        ■ SUBORDER ANTHROPOIDEA: monkeys, apes, men
          ■ SUPERFAMILY CEBOIDEA: New World monkeys   **138**
          ■ SUPERFAMILY CERCOPITHECOIDEA: Old World
            monkeys   **138**
          ■ SUPERFAMILY HOMINOIDEA: apes, men
            ■ FAMILY HYLOBATIDAE: gibbons, *Pliopithecus*
              138, 148
            ■ *Family Oreopithecidae: Oreopithecus*   **138, 148**
            ■ FAMILY PONGIDAE: *Dryopithecus*, orangutan,
              chimpanzee, gorilla   **139, 145, 148**
            ■ FAMILY HOMINIDAE: *Ramapithecus*,
              Australopithecus, Homo   **139, 146, 148,**
              149, 151, 152, 154

"Placental Mammals"

suggestions for further reading

GENERAL

DOTT, R. H., JR. and R. L. BATTEN, 1971, *Evolution of the Earth*. New York: McGraw-Hill.

KEETON, W. T., 1972, *Biological Science*. New York: W. W. Norton.

RAUP, D. M. and S. M. STANLEY, 1971, *Principles of Paleontology*. San Francisco: W. H. Freeman.

ROMER, A. S., 1968, *The Procession of Life*. Cleveland: World Publishing Co.

STOKES, W. L., 1973, *Essentials of Earth History*. Englewood Cliffs, New Jersey: Prentice-Hall.

CHAPTER 1
THE BEGINNINGS OF LIFE

CALVIN, M., 1975, "Chemical Evolution," *American Scientist*, vol. 63, p. 169–177.

CLOUD, P., 1974, "Evolution of Ecosystems," *American Scientist*, vol. 62, p. 54–66.

KNOLL, A. H. and E. S. BARGHOORN, 1975, "Precambrian Eucaryotic Organisms: A Reassessment of the Evidence," *Science*, vol. 190, p. 52–54.

MILLER, S. L. and L. E. ORGEL, 1974, *The Origins of Life on Earth*. Englewood Cliffs, New Jersey: Prentice-Hall.

SCHOPF, J. W., 1975, "Precambrian Paleobiology: Problems and Perspectives," *Annual Review of Earth and Planetary Sciences*, vol. 3, p. 213–249.

CHAPTER 2
THE DIVERSIFICATION OF LIFE

DEBEER, G. R., 1964, *Atlas of Evolution*. London: Nelson.

SAVAGE, J. M., 1969, *Evolution*. New York: Holt, Rinehart & Winston.

SIMPSON, G. G., 1967, *The Meaning of Evolution*. New Haven: Yale University Press.

STEBBINS, G. L., 1977, *Processes of Organic Evolution*. Englewood Cliffs, New Jersey: Prentice-Hall.

CHAPTER 3
LIFE IN THE SEA

BARRINGTON, E. J. W., 1967, *Invertebrate Structure and Function*. Boston: Houghton Mifflin.

BEERBOWER, J. R., 1968, *Search for the Past*. Englewood Cliffs, New Jersey: Prentice-Hall.

BUCHSBAUM, R. M., 1972, *Animals Without Backbones*. Chicago: University of Chicago Press.

FRIEDRICH, H., 1970, *Marine Biology*. Seattle: University of Washington Press.

CHAPTER 4
THE TRANSITION TO LAND

BALDWIN, E., 1964, *An Introduction to Comparative Biochemistry*. Cambridge: Cambridge University Press.

CARTER, G. S., 1967, *Structure and Habit in Vertebrate Evolution*. Seattle: University of Washington Press.

MARSHALL, N. B., 1966, *The Life of Fishes*. New York: Universe Books.

CHAPTER 5
LAND PLANTS

BANKS, H. P., 1970, *Evolution and Plants of the Past*. Belmont, California: Wadsworth.

BOLD, H. C., 1977, *The Plant Kingdom*. Englewood Cliffs, New Jersey: Prentice-Hall.

CORNER, H. J. H., 1964, *The Life of Plants*. Cleveland: World Publishing Co.

DELEVORYAS, T., 1963, *Morphology and Evolution of Fossil Plants*. New York: Holt, Rinehart & Winston.

CHAPTER 6
REPTILES AND MAMMALS

BAKKER, R. T., 1975, "Dinosaur Renaissance," *Scientific American*, vol. 232, no. 4, p. 58–78.

BOURLIÈRE, F., 1964, *The Natural History of Mammals*. New York: Alfred A. Knopf.

KURTÉN, B., 1968, *The Age of Dinosaurs,* New York: McGraw-Hill.

ROMER, A. S., 1966, *Vertebrate Paleontology,* Chicago: University of Chicago Press.

STAHL, B. J., 1974, *Vertebrate History: Problems in Evolution*. New York: McGraw-Hill.

PILBEAM, D. R., 1972, *The Ascent of Man*. New York: Macmillan.

ROSEN, S. I., 1974, *Introduction to the Primates*. Englewood Cliffs, New Jersey: Prentice-Hall.

SIMONS, E. L., 1972, *Primate Evolution*. New York: Macmillan.

CHAPTER 7
MAN

LEAKEY, R. E., 1976, "Hominids in Africa," *American Scientist,* vol. 64, p. 174–178.

index

Acorn worms, 77
Acritarchs, 56–58
Adaptive divergencies, 41–42
Age of Metals, 153
Agnatha, 79, 80–81
Agriculture, 153
Algae, 44, 47, 55, 70, 89
    See also Procaryotes
Alleles, 34–36
    individual variation and, 36–37
    population genetics and, 37–39
Allopatric species, 41
Amino acids, 5
    in primitive earth experiments, 7–8
Ammonia, 6–7, 8
Amphibians, 78
    development of, 85–88
    reptiles and, 109–111
Angiospermophyta, 100
    development of, 105–107
Ankylosaurs, 121
Annelida, 24, 52, 61, 70, 77
    ancestors of, 22
    fossils of, 55, 60, 62
Anthropoidea, 140, 143
Apes, 137
    pre-Pleistocene, 143–146
Arachnids, 72–74
Aragonite, 52, 54
Archeocyathids, 61
Armadillos, 134, 136
Armored fish, 80–81
Arthrodires, 81
Arthropods, 54, 60, 70, 72, 77
Artiodactyla, 135
Atmosphere of the earth, 5–6, 26
    extinction and, 68
Australia, 131, 134
Australopithecus, 148–150
Australopithecus africanus, 149
Australopithecus robustus, 149
Autotrophs, 9–10

Baboons, 145
Bacteria, 10, 11–12, 47, 54, 70
    Precambrian fossils of, 16–17
Barghoorn, E. S., 16
Barriers of speciation, 41
Bats, 135
Bears, 135
Benthos (bottom-living organisms), 45,
    47–49
    invertebrate, 73
    invertebrate adaptation, 49–52
    fossils of, 55–56
    plants, 56
Binomial nomenclature, 29
Bipeds, 116, 145–147, 150
Birds, 75, 78, 117, 126
    fossils of, 108–109
Bitter Springs Formation fossils, 18–19
Bivalves, 52, 60, 66
Blue-green algae, *see* Procaryotes
Bony fish (osteichthyes), 79, 82
    development of, 83–85
Brachiopods, 25, 52, 54, 60, 65
Brain, 128, 137
    primate, 147–152
Brown algae (phaeophyta), 54, 56
Bryophyta, *see* Mosses
Bryozoans, 52, 59, 60, 66, 72
Budding, 26
Burgess shale fauna, 62–64

Cacti, 107
Calcite, 52, 54
Calcium carbonate, 52–54
Calcium phosphate, 52–54
Cambrian Period
    early paleozoic invertebrates of, 59–65
    marine invertebrate extinctions in, 65–68
    marine life development during, 45–56
    marine plants during, 56–58
    rise of shell-bearing metazoans during,
        24–27

Carbohydrates, 8, 47, 70
  chemical fossil traces of, 15
  as a component of life, 4–5
  photosynthesis and, 9–10
Carbon, 5–6, 8, 15
  traces of Precambrian compounds of,
    13–15
Carbon dioxide, 6, 8, 26, 47
  structural material synthesis from, 9
Carboniferous Period, 73, 74, 86–87, 110
  plants of, 92, 97, 101, 104
Carnivores, 81, 117–121, 135
Carnivores-scavengers, 49
Cartilage, 83
Casts, 55
Cats, 135
Cave paintings, 153
Ceboidea, 143
Cedar trees, 104
Cenozoic Era, 43, 73, 107
  mammals of, 131–135, 137
  primates of, 140
Cephalochordata, 75–77
Cephalopods, 47, 60, 67
Cercopithecoidea, 143
Charophyta, 70
Chemical model of origin of life, 4–8
  components of, 4–5
  early earth environment and, 5–6
  experiments in, 6–8
Chemical traces of early life, 13–15
Cherts, 15–16
Chewing, 128
Chimpanzee, 143, 144
Chlorophyll, 71
Chlorophyta, 70
Chloroplasts, 19–21
Chordata, 70, 75
  invertebrate, 75–77
Chromosomes, 11, 21
  genetic theory and, 33–37
Chrysophyta, 70
Clams, 25, 51, 52, 72, 73
Classes, 29, 42
Classification systems, 29
Claws, 116, 140
Climate change, 26, 68, 147
Coccolithophorids, 47, 55, 56
Coelenterata, 22, 24, 60, 62, 72
Compressions, 90
Conifers, 70, 89, 100, 101, 104–105
Convergence, 43

Copepods, 47
Corals, 52, 60, 65–66
Cotylosaurs, 112–115, 116
Crabs, 25
Credonts, 135–136
Cretaceous Period, 43, 56, 67, 83, 86, 112
  plants of, 90, 100
  reptiles of, 116, 117
Crinoids, 52, 65
Crocodiles, 117
Crossopterygians, 86
Crustaceans, 51–52, 55, 61, 66, 72, 73
Cultivation of plants, 153
Culture, evolution of human, 153–154
Cuticle, 71
Cycads, 100, 101, 103–104

Darwin, Charles, 30–32
Deposit feeders, 49–51, 80
Devonian Period, 44, 56, 71, 73
  amphibian evolution during, 85–88
  fish evolution during, 79–82
  plant life of, 92, 95
Diatoms, 47, 52, 55, 56
"Dilute organic soup," 8
Dinoflagellates (pyrrophyta), 47, 54, 55,
  56
Dinosaurs, 112, 116–123
Diploid cells, 34–35
DNA (deoxyribonucleic acid), 33
  mutation and, 37
Dogs, 135
Domestication of animals, 153
Dominant alleles, 36
Drosophila, 36
Dryopithecines, 144–146

Earth, 3
  atmosphere of, 5–6, 26, 68
Echinodermata, 60, 77
Echinoids, 66
Ecologic barriers, 41
Ediacara fauna, 21–24
  Cambrian development of, 24–27
Egg cells, 26, 33, 94
Egg laying mammals, 131
Eggs, shelled, 109, 129
Elephants, 135
Embryonic development, 128
Energy source, origins of life and, 8, 10–11
Environmental variation, 36–37, 42
  extinctions and, 43

Enzymes, 5
Eocene Epoch, 43, 134
Epifaunal animals, 49–51
Eucaryotes, 11–12
  Precambrian fossil traces of, 19–27
Evolution, organic, 28
  Darwin and, 30–32
  fossil records and, 42–43
  genetic theory and, 33–37
  human, 137
  modern theories of, 37–42
Evolutionary radiation, 43, 65
  of fish, 79, 83
  of mammals, 134
  of plants, 107
  of reptiles, 110, 116
External skeleton, 54, 77
Extinctions, 43
  dinosaur, 123
  glaciation and, 147
  of mammals, 135–136
  of marine invertebrates, 65–68
  of vascular plants, 101
  of vertebrates, 82
Eyes, primate, 140

Families, 29, 43
Fats, 10
  as a component of life, 4–5
Fayum deposits (Egypt), 144
Feathers, 126
Ferns, 47, 70, 93, 97, 101, 105, 106
Fertilization, 34, 69, 95, 100, 105
  recombination during, 37
Fig Tree Fossil group, 17–18
Filter-feeders, 49–51
Fins, 83–85
Fir trees, 104
Fish, 47, 56, 75
  evolutionary history of, 78–85
Flight, 73
  mammal, 135
  reptile, 117
Flowering plants, 70, 90
  development of, 105–107
Fluid-transport system of plants, 71, 89
Flying bats, 134, 135
Flying reptiles, 117
Food-gathering organs, 24
  of agnaths, 80
  of benthonic invertebrates, 49–51
  of trilobites, 62

Foraminiferans, 47
Fossils, 1, 3, 13, 71
  amphibian, 86
  of benthonic invertebrates, 49
  bird, 108–109
  evolutionary patterns in, 42–43
  fish, 79
  insect, 74
  land plant, 90–91, 107
  mammal, 128
  marine, 52–56
  primate, 144–145
  procaryotes, 15–19
  reptile, 114
  of terrestrial invertebrates, 73
Fresh water phyla, 70
Frogs, 87, 111
Fruit, 105
Fungi (mycophyta), 47, 54

Gametes, 33–36, 69, 71
Gametophyte, 94, 100
Gastropods, 52, 72, 73
Genes, 33–37
Genetic equilibrium, 37–38
Genetics, 33–37
  population, 37–39
Genus, 29, 42
Geographic isolation, 40–42
Gibbon, 143, 146
Gill slits, 75, 77
Ginkgoes, 100, 101, 103–104
Giraffes, 135
Glaciation, 146–147
Glaessner, M. F., 22
Gorilla, 143, 144
Gunflint Chert fossils, 15–17
Gymnosperms, 89–90, 100–104, 106

Hagfish, 91
Hair, 126
Haldane, J. B. S., 8
Hands, primate, 140
Haploid cells, 35
Hardy-Weinberg law, 37–38
Hardy, G. H., 37
Hearing, 128
Heart, 126
Hedgehogs, 134
Hemichordata, 77
Herbivores, 49–51, 81, 117–121
  placental, 134

Heredity, 33
  individual variation and, 36–37
Heterotrophs, 11
Heterozygous individuals, 35–36
Hominidae, 137, 143, 148
Homo erectus, 148, 150–151
Homologous chromosome pairs, 34
Homo sapiens, 137, 147, 148, 150–151
Homozygous individuals, 35–36
Horned dinosaurs, 121
Horses, 135
Hybridization, 41
Hydrogen, 5–6, 7–8

Ice Age, 146
Ichthyosaurs, 115
Incisors, 128
Individual variation, 36–37
Industrial melanism, 39
Infaunal animals, 49–51
Inheritance of variability, 33
  genetic theory and, 36–37
Insectivora, 134, 138
Insectivorous plants, 107
Insects, 72–74, 117
Intelligence, primate, 147, 151
Interbreeding populations, 37–39
Internal skeletons, 54, 77
Invertebrates, 44
  chordate, 75–77
  early Paleozoic, 59–65
  extinctions of marine, 65–68
  terrestrial, 72–75
Isolating mechanisms, 41

Java Man, 150
Jaw development, 81, 83, 126
Jawless fish (agnaths), 80–81
Jellyfish, 47, 55, 61
Jurassic Period, 56, 66, 87, 101, 106
  reptiles of, 112, 116, 117–121

Kangaroos, 133
Koala bears, 134

Labyrinthodonts, 87–88, 111, 112
Lamarck, Jean Baptiste de, 29
Lamarckism, 30–31
Lampreys, 81
Lancelets, 75–77
Latimeria, 86
Leaves, 71, 89
Legs, 85
Lemurs, 140

Life:
  components of, 4–5
  earliest multicellular, 21–24
  energy source for, 10–11
  origins of, 3–8
  structural materials of, 9–10
  transition to land of, 69–88
  See also Evolution, organic
Lightning, 7, 8
Limb development, 85–86, 129
  thecodont reptile, 116–117
Linneaus, Carolus, 29
Liverworts, 70–71, 89
Lizards, 114
Lobe-finned bony fish, 85
Locomotion, 72, 85, 86, 129
  biped, 116, 150
Lorises, 142
Lungfish, 85–86
Lungs, 72, 85, 126
Lycopsida, 92, 94, 97, 101, 106

Madagascar, 140–142
Major adaptive divergencies, 41–42
Mammals, 43, 57, 75, 78, 117
  development of, 126–136
  reptilelike, 112–113, 128–129
  See also Man
Mammary glands, 128
Mammoths, 136, 147
Man, 137–154
  cultural evolution of, 153–154
  Homo erectus and, 150–152
  Pleistocene expansion of, 146–150
  pre-Pleistocene monkeys, apes and, 143–146
  primate origins, adaptations and, 138–142
Marine life
  early Paleozoic, 59–65
  fossilization of, 52–56
  invertebrate extinctions of, 65–68
  land transition of, 69–70
  life modes of, 45–52
  vertebrate, 75
Marine plants, 55, 56–58
  transition to land of, 70–72
Marine reptiles, 56, 115
Marsupials, 132–133, 134
Mastodons, 136, 147
Meiosis, 33–35
  recombination during, 37
Mendel, Gregor, 33

Mesozoic Era, 43, 65, 67, 111
  mammals of, 128–130
  plants of, 90, 100
Metal tools, 153–154
Metaphytes, 19–21
Metazoans, 19–21, 42
  theories of development of, 25–27
Meteorites, 5, 8
Methane, 6–8
Milk, 128
Miller, S. L., 6–7
Miocene Epoch, 135
  primates of, 144–146
Mistletoe, 107
Mitochondria, 19–21
Mitosis, 33–34
Mobility, benthonic, 49–51
Molars, 128
Mold fossils, 55
Moles, 134
Molluscs, 51, 55, 60–61, 66, 70, 72
Monkeys, 135, 137
  pre-Pleistocene, 143–146
Monophyletic, 43
Monotremes, 131
Mosses (bryophyta), 54, 57, 70–71, 89
Moths, industrial melanism in, 38–39
Multicellular fossil life, earliest, 21–24
Muscles, 24, 77
Muscular foot, 85
Mutation, 36–37, 43, 44

Natural selection, 30–32, 42, 44
  population genetics and, 37–39
  speciation and, 41
Nautilus, 47, 67
Neanderthal Man, 151
Nekton (swimming), 45, 47
Neolithic Age, 153
Nerve cells, 24
Nervous system, 77, 128
New Stone Age, 153
Newts, 87
New World monkeys, 143–144
Nitrogen, 5–6, 8, 26
  autotrophs use of, 9–10
Nitrogen-fixers, 10
Nonphotosynthetic autotrophs, 10
Notochord, 75–77
Nucleic acids, 21
  as a component of life, 4–5
  DNA, 33

Nucleotide bases, 5, 8
Nucleus, cell, 11, 19, 33
  in eucaryotes, 21
Nuclides, 15

Octopi, 47, 67
Old Stone Age, 153
Olduvai Gorge (Tanzania), 149
Old World monkeys, 143–144
Oligocene Epoch, 135
  primates of, 143–144
Oparin, A. I., 8
Opossums, 133
Opposable thumb, 140
Orangutan, 143, 144
Orders, 29, 42
Ordovician Period, 44, 56, 59, 61, 65, 75
  fish fossils of, 79, 80
Oreopithecus, 146
Organic evolution, *see* Evolution, organic
Origin of Species, The (Darwin), 31–32
Origins of life, 3–8
Ornithischia, 117
Ornithopods, 121
Oxygen, 5–6, 8, 68
  development of metazoans and, 26–27
  life transition to land and, 69, 72, 85

Paleocene Epoch, 134
  mammals of, 138
Paleolithic Age, 153
Paleontology, 1
Paleozoic Era, 43, 57, 67, 82
  invertebrate life in, 59–65
  plant life of, 89, 92–99, 100, 106
Parallelism, 43
Parasites, 16, 107
Parasitism, 81
Pathogenic organisms, 68
Perissodactyla, 135
Permian Period, 43, 65–66, 74, 82, 88
  plant life of, 92, 97, 101
  reptile life of, 110
Petals, 105
Petrified wood, 90–91
Phanerozoic, 13, 25
Photosynthesis, 9, 26, 69, 70–71, 89, 97
  age of, 17, 18
  chloroplast development and, 19–21
  of plankton, 47
Phyla, 29, 42, 44
  fossil record of, 42–43

Physiological barriers, 41
Phytoplankton, 45, 55
Pine trees, 104
Placental mammals, 132–133, 134–137, 138
Placodermi, 79, 81–82
Plankton (floating), 45, 47, 80
  fossils of, 55–56
  plants, 56
Plants, 9–10, 59
  cultivation of, 153
  land, 88–107
  marine, 55, 56–58
  polyploidy among, 41
  Precambrian, 44
  rise during Cretaceous Period of, 43
  transition to land of, 70–72
Platypus, duck-billed, 131
Pleistocene expansion of man, 146–153
  glaciation and, 146–147
Plesiosaurs, 115
Pliocene Epoch, 144–146
Pliopithecus, 146
Pollen, 89–90, 91, 100
Pollination, 105
Polyphyletic, 43
Polyploidy, 39, 41
Polytypic species, 40
Population genetics, 37–39
Porifera, 60
Porpoises, 47, 135
Pottery, 153
Precambrian fossils, 3, 15–19
  eucaryotes, 19–27
  procaryotes, 15–19
Precambrian life, 13–19
  chemical traces of, 13–15
Premonkeys, 137, 140–142
Primates, 135, 137
  origins and adaptations of, 138–140
  See also Man
Primitive earth experiments, 6–8
Proboscidea, 135
Procaryotes (blue-green algae), 11–12, 26–27, 56
  Precambrian fossil evidence of, 13, 15–19
Prosimians, 140–142
Proteins, 9–10
  as a component of life, 4–5
Protozoans, 72
Psilopsida, 92, 94, 95
Pteridospermophyta, 101
Pteropods, 47
Pteropsida, 92
Pterosaurs, 117

Radiation, 36, 68
Radiolarians, 47, 52
Ramapithecus, 146, 147
Ray-finned bony fish, 85
Recessive alleles, 36
Recombination, 37, 42, 44
Recrystallization, 54
Replacement, 54
Reproductive system, plant, 89–90
  seed-bearing, 100
  seedless, 94
Reptiles, 75, 78, 86
  development of, 108–125
Respiration, 72, 85
Rhinoceroses, 135
Rhynie Chert, 90, 95
Rodentia, 134, 135, 140
Roots, 71, 89, 97

Sabre-toothed cats, 136, 147
Salamanders, 87, 111
Salt, 70, 81
Sarcodina, 60, 61, 66, 70
Saurischia, 117
Sauropods, 121
Schizomycophyta, *see* Bacteria
Seals, 47, 135
Sea urchins, 77
Seaweed, 47
Sedimentation, fossilization and, 56
Sediment feeders, 49–51
Seed-bearing plants, 47, 100–104
Seed ferns, 100, 101
Seedless Paleozoic plants, 92–99
Seeds, 89–90
  of angiospermophyta, 105
Self-reproduction, 8
Sexual reproduction, 24
  of amphibians, 86
  genetic theory and, 33–37
  life transition to land and, 69, 85
  of mammals, 126, 131–133
  of reptiles, 109–110
  rise of metazoans and, 26
  species differentiation and, 39–41
  See also Reproduction system, plant
Sharks (chondrichthyes), 79, 80, 82
  development of, 83–85
Shell bearing marine life, 24–25
  fossil traces of, 52–56
  late Cambrian rise of, 42
Shelled egg, 109, 129
Shrews, 134
Siblifng species, 40

Sight, primate, 140
Silica, 52–54
Silicification, 54
Silurian Period, 44, 69, 79, 80, 89
    plant life of, 92, 95
Skeletons, 24, 77
    of birds, 126
    of invertebrate phyla, 60
    marine fossils and, 52–56
    of primates, 138–140
    of reptiles, 109–110
    shark, 83
Sloths, 134, 136, 147
Smell, sense of, 128
Snails, 25, 51, 52, 60, 72
Snakes, 114
Solar radiation, 8, 9, 47, 71
Sound, 128
Sound waves, 8
South America, 134
Speciation, 39–41
Species, 29, 40
Sperm cells, 26, 33, 95
Spermopsida, 92, 94, 100
Sphenopsida, 92, 94, 97, 101, 106
Spiny anteater, 131
Sponges, 25, 51, 52, 61, 66, 72
Spontaneous generation, 29
Spores, 91, 94
Sporophyte, 94, 100
Squid, 47, 67
Squirrels, 140
Starfish, 25, 60, 77
Stegosaurs, 121
Stem reptiles, 112
Stereoscopic vision, 140
Strategies of living systems, 9–12
Stromatolites, 15, 18
Structural barriers, 41
Structural materials of life, 9–10
Subspecies, 40
Sulfur, 10
Sunlight, 5, 10
Suspension-feeders, 80
Swimming, 83
Sympatric species, 41

Tarsiers, 142
Teeth, 54, 55, 126–127
Temperature, mutation and, 36
Terrestrial life transition, 69–88
    amphibian, 85–88
Thecodonts, 116–117
Theropods, 121

Thumb, opposable, 140
Toads, 87
Tool use, 137, 140, 147, 150
    human cultural evolution and, 153
Tracheae, 72
Tracheophyta (vascular plants), 54, 70–71, 89, 92
Transformism, 29–30
Transmutation, 30–31
Tree-dwellers, 138–140, 144–145
Trees, 90, 103, 106–107
    seedless, 97
Triassic Period, 65–66, 87–88, 106
    reptiles of, 111, 115, 116–121
Trilobites, 25, 60, 62, 65
Turtles, 47, 114
Tyler, S. A., 16
Tyrannosaurus, 121

Ungulates, archaic, 135–136
Unicellular green plants, 47
Urochordata, 77

Variability, inheritance of, 33
    genetic theory and, 36–37
Vascular plant systems, 71–72, 89, 95
Vertebrae, 75
Vertebrates
    amphibians, 85–88
    fish evolution and, 78–85
    origin of, 75–77
    reptiles, 108–125
Vision, primate, 140
Volcanic gases, 6, 26

Walcott, C. D., 62
Wallabies, 134
Wallace, Alfred Russel, 31
Warm-blooded, definition of, 126
Water, 26
    as a component of life, 4–5
    as "dilute organic soup," 8
    structural material synthesis from, 9–10
    transition to land of marine life and, 69, 72, 85
Weinberg, G., 37
Whales, 47, 134
Wings, 116–117
Wombats, 134
Worms, 51, 72
Writing, 154

Zooplankton, 45, 47

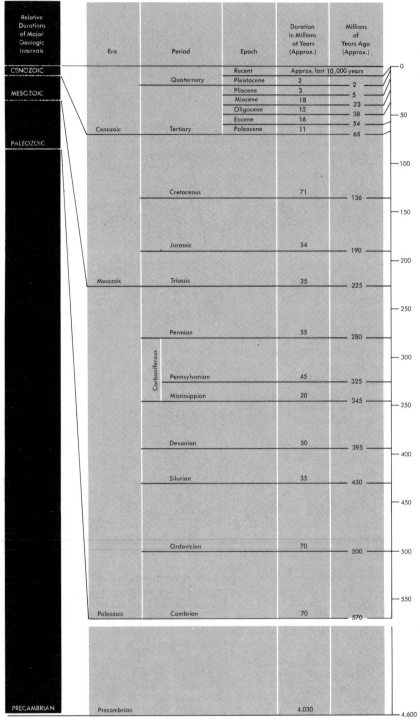

Relative Durations of Major Geologic Intervals	Era	Period	Epoch	Duration in Millions of Years (Approx.)	Millions of Years Ago (Approx.)
CENOZOIC			Recent	Approx. last 10,000 years	0
		Quaternary	Pleistocene	2	2
MESOZOIC			Pliocene	3	5
			Miocene	18	23
			Oligocene	15	38
PALEOZOIC	Cenozoic	Tertiary	Eocene	16	54
			Paleocene	11	65
		Cretaceous		71	136
		Jurassic		54	190
	Mesozoic	Triassic		35	225
		Permian		55	280
		Carboniferous Pennsylvanian		45	325
		Mississippian		20	345
		Devonian		50	395
		Silurian		35	430
		Ordovician		70	500
	Paleozoic	Cambrian		70	570
PRECAMBRIAN	Precambrian			4,030	4,600

Formation of Earth's crust about 4,600 million years ago

Millions of Years